Connected

I0446520

The Psychology of Technology

Luis Tejada

Index

Introduction

In a world where technology has become omnipresent, shaping every aspect of our daily lives, a fascinating and sometimes unexplored intersection emerges between digital machinery and the complexity of the human mind. This book, "Connected: Exploring the Psychology of Technology," delves into the depths of this intricate relationship, unraveling the invisible threads that connect our psychology with technological advances that, often unnoticed, shape our reality.

We live in an era where virtual connections compete with face-to-face interactions, where the line between reality and simulation blurs with each technological advancement. In this digital space, our psychology navigates unknown territories, facing unique challenges and experiencing new forms of social interaction, self-image, and emotional well-being.

As our devices become smarter, our lives more connected, and our experiences more digitized, a series of crucial questions arise. How does constant exposure to technology affect our mental health? What role do social networks play in shaping our identity and self-perception? To what extent does artificial intelligence understand and respond to our emotions?

This book not only seeks to answer these questions but also invites readers to reflect on the changing nature of our relationship with technology. Through a multidisciplinary approach spanning psychology, sociology, ethics, and neuroscience, "Connected" ventures into the very heart of the human experience in the digital age.

Exploring topics ranging from social media addiction and the impact of blue light on sleep to online authenticity and technological stress, this book aims to shed light on the complexities of modern life. As we move towards a future where technology will continue to evolve rapidly, understanding how it affects our psychology becomes essential to forge a harmonious balance between innovation and our emotional well-being.

"Connected: Exploring the Psychology of Technology" is more than an examination of interactions between mind and machine; it is an invitation to reflect on our role in creating a future where technology not only connects us to the world but also to ourselves.

2.Social Media Addiction

In the era of instant connectivity, social networks have woven a virtual web spanning the globe, allowing us to share, communicate, and explore the world from the palm of our hand. However, behind this apparent marvel of interconnection lies a growing phenomenon that has captured the attention of psychologists and mental health experts: social media addiction.

This addiction often finds its roots in the constant search for social validation. The prospect of receiving likes, comments, and followers has become a digital social currency that fuels the innate human desire to be accepted and recognized. However, this instant validation can turn into a trap, generating a compulsive need for attention and approval.

In the vast landscape of social networks, the relentless pursuit of social validation has become a central element driving addiction. This phenomenon is rooted in the deep human need to be acknowledged, accepted, and valued, and social platforms have provided a unique stage where these aspirations find their digital expression.

Obtaining "likes," comments, and followers has emerged as a digital social currency, where each positive interaction becomes an instant validation of one's worth. In an increasingly interconnected world, this social validation has become more tangible and accessible than ever. However, what begins as a natural longing for connection and recognition can gradually transform into a complex trap.

This operates insidiously. Initially, receiving approval in the form of likes and comments provides a gratifying sensation, releasing that surge of dopamine linked to pleasure and reward. This positive stimulus reinforces the behavior, creating a direct association between social media activity and emotional gratification. The human mind, eager for instant satisfaction, falls into the trap of validation as a constant source of emotional well-being.

Over time, this instant validation becomes an addictive fuel. The need for attention and approval becomes compulsive, generating a constant search for positive online interactions. Each "like" becomes a measure of self-esteem, and the absence of these validations can trigger feelings of insecurity and anxiety. At this point, social validation ceases to be a conscious choice and becomes an urgent need, a driving force shaping behaviors and decisions in the digital world.

The trap of social validation not only affects self-perception but can also distort reality. The need to present a constantly positive image can lead to the creation of an idealized version of life, triggering a cycle of constant comparison with others. The gap between real life and digital life can widen, fostering a sense of disconnection between online and offline identity.

The act of thumb-scrolling to update the feed or receive a notification activates the same reward centers in the brain as addictive substances. The release of dopamine, a neurotransmitter associated with pleasure and reward, creates a sense of immediate satisfaction. This neurological response reinforces the behavior and contributes to the addictive cycle of constantly checking social media.

This act leads to the release of dopamine. This neurotransmitter, known as the messenger of pleasure and reward in the brain, becomes a central player in the stage of social media addiction, contributing to the creation of a deep and sometimes dangerous relationship between the human mind and technology.

When we interact with social media, especially when receiving positive notifications such as a "like" or a comment, reward centers in the brain, like the nucleus accumbens, are notably activated. This process is similar to what occurs with addictive substances like cocaine or nicotine. Dopamine, released in response to these digital experiences, generates a feeling of pleasure and instant gratification, creating a neurochemical bond between the use of social media and the sense of well-being.

The release of dopamine is not only associated with pleasure but also plays a crucial role in behavior reinforcement. When we experience pleasure as a result of a specific action, the brain associates that action with the pleasurable sensation and reinforces the likelihood of repeating that behavior in the future. In summary, in the context of social media, every time the feed is updated and positive notifications are found, the connection between interacting and gratification is reinforced, contributing to the formation of addictive behavioral patterns.

The constant pursuit of positive interactions becomes a routine, fueled by the hope of obtaining the next dose of dopamine. This cycle, similar to other addictions, can generate a compulsive need to check and engage in social media as the mind seeks to replicate the pleasurable sensation associated with dopamine release.

Over time, the brain may become desensitized to dopamine, meaning that an increasingly greater stimulus is needed to experience the same feeling of pleasure. This phenomenon may explain why some individuals get trapped in cycles of increasingly frequent social media use, constantly seeking a higher level of digital gratification.

Understanding the connection between dopamine and social media addiction is crucial for developing effective strategies to break the cycle. Establishing time limits, practicing regular digital disconnection, and fostering awareness of the relationship between social media use and dopamine release are essential steps to regain control over the mind-technology connection.

Social media also encourages constant comparison. Carefully edited and filtered lives presented on platforms like Instagram and Facebook can evoke feelings of inadequacy and anxiety. Constant exposure to seemingly superior achievements and experiences can fuel an endless quest for comparison, triggering a cycle of dissatisfaction and the need for validation.

Within the vibrant world of social media, constant comparison stands as a persistent shadow, lurking on the periphery of every scroll and click. Platforms where others' lives unfold in selected and carefully filtered images become the stage where a silent but fierce battle between reality and digital representation unfolds.

The very essence of social media, based on the selection and presentation of highlighted moments, often creates an idealized version of life. Perfectly composed photos and carefully curated narratives construct a storyline that, while visually appealing, rarely reflects the complexity and struggles of everyday existence. This gap between reality and representation contributes to constant comparison, as users measure themselves through the distorted prism of others' digital lives.

Constant comparison fosters a breeding ground for feelings of inadequacy and anxiety. Bombarded with images of idealized achievements, experiences, and appearances, users may feel that their own lives fall short. This sense of inadequacy intensifies with each scroll, creating a chain of negative thoughts undermining self-esteem and contributing to the feeling that one's own life is inferior.

Constant comparison also gives rise to a phenomenon known as the social envy effect. Constantly observing the seemingly perfect lives of others can generate envy and resentment, emotions that, in the long run, can affect interpersonal relationships and mental health. Social envy becomes a toxic byproduct of constant comparison, contaminating the digital experience and undermining the ability to enjoy one's own achievements and experiences.

It also fuels the incessant pursuit of validation. Users may feel the constant need to compete in the game of digital representation, seeking to obtain the same approval and recognition they perceive in others' lives. This perpetual cycle of comparison and validation-seeking contributes to the addiction trap, as the mind constantly seeks to close the perceived gap between its own reality and the idealized digital representation.

Breaking the cycle of constant comparison requires a conscious effort towards self-acceptance and appreciation of authenticity. Fostering a mindset of gratitude, limiting exposure time to idealized content, and remembering that social media is only a fraction of reality can help mitigate the negative effects of constant comparison.

Excessive use of social media has been linked to mental health problems such as depression, anxiety, and low self-esteem. Constant connection to others' lives can create relentless pressure to maintain a flawless image, contributing to a sense of falling short. Additionally, social media addiction can lead to offline social isolation, creating a paradox in which virtual connectivity coexists with real loneliness.

The seemingly endless universe of social media, with its constant flow of images and updates, has triggered a silent but pervasive phenomenon: the impact on mental health. As society plunges deeper into digital waters, a complex landscape is revealed where excessive use of social media can leave deep psychological scars.

Numerous studies have pointed out the connection between excessive social media use and mental health problems, including depression and anxiety. Constant exposure to seemingly perfect lives can generate negative comparisons and feelings of inadequacy. The pressure to maintain a flawless image contributes to anxiety, while the gap between real life and digital representation can fuel depression.

The incessant pursuit of validation and constant comparison with others' digitally idealized lives can undermine self-esteem. Users may experience a persistent sense of falling short as they measure themselves against the artificially elevated standards of social media. This low self-esteem can manifest in daily life, affecting social interactions and the perception of one's own worth.

The constant pressure to maintain a flawless online image can become a significant emotional burden. The need to portray a perfect life can generate additional stress, leading to the creation of a digital facade that hides real struggles and challenges. This carefully crafted representation can deplete mental and emotional resources, contributing to mental fatigue.

In a surprising paradox, virtual connectivity can coexist with real loneliness. Social media addiction, instead of alleviating loneliness, often exacerbates it. Offline social isolation can be a direct consequence of constant online connection. The paradox lies in the fact that, despite the extensive digital network, individuals may feel lonely and disconnected from the reality around them.

Developing awareness of negative thought patterns, establishing healthy limits on social media use, and seeking offline support are fundamental steps to break the cycle of harmful virtual connectivity.

Recognizing and addressing social media addiction is essential to preserve mental health. Strategies such as setting time limits, practicing regular digital disconnection, and fostering awareness of the psychological impact of social media are crucial steps toward a healthier relationship with technology.

A fundamental strategy for healthy disconnection is to establish clear and realistic time limits for social media use. Defining specific periods of the day dedicated to online interaction and limiting the total daily time can help avoid constant immersion in the digital world. This practice provides an essential balance, allowing moments of connection without allowing social media to dominate daily life.

Scheduling regular periods of digital disconnection is essential for maintaining mental health. Establishing days or moments when notifications are turned off and active participation in social media is avoided can help reduce dependence and restore perspective. Regular digital disconnection provides an opportunity to connect more fully with the offline environment, fostering more meaningful interpersonal relationships and reducing the constant pressure to maintain an online presence.

Awareness of the psychological impact of social media is essential for addressing addiction. Taking the time to reflect on how the use of these platforms affects self-esteem, emotions, and relationships can be a significant first step. Maintaining a personal journal or engaging in reflective discussions with friends and loved ones can help uncover harmful behavior patterns and promote a more conscious connection with technology.

Designating specific areas or times where technology is prohibited can contribute to healthy disconnection. Establishing technology-free spaces at home or during specific activities, such as meals or family time, creates opportunities for mindfulness and genuine connection with the environment and people around.

Setting offline goals can be an effective strategy to shift focus and reduce dependence on social media. These goals may include activities such as reading books, practicing sports, learning new skills, or engaging in social events without the constant need to document it online. Establishing offline goals provides a sense of accomplishment independent of digital validation.

When social media addiction significantly impacts mental health, seeking professional support is crucial. Psychologists, therapists, or counselors can provide tailored tools and strategies to address digital dependence and work on building a healthier relationship with technology.

These strategies not only help counteract social media addiction but also promote a more mindful connection with the digital world. By implementing these practices, individuals can preserve their mental health and build a balanced and healthy relationship with technology in the digital age.

3.FOMO (Fear of Missing Out)

In the vast ocean of social media, the tide of FOMO, or Fear of Missing Out, flows strongly. This psychological phenomenon, amplified by constant online connectivity, represents a palpable anxiety stemming from the perception that others are living more exciting, interesting, or meaningful experiences.

Social media, platforms designed for sharing and connecting, have also become digital stages where carefully crafted and idealized versions of life are presented. Selective images and compelling narratives can create the perception that others' lives are filled with thrilling moments and unforgettable experiences. This digital environment of constructing attractive realities acts as the perfect breeding ground for FOMO, generating the feeling that others are participating in significant events from which one is absent.

In the virtual fabric of social media, an intriguing and sometimes deceptive phenomenon unfolds: the digital construction of attractive realities. In this digital scenario, platforms originally designed for sharing moments and connecting with others have transformed into stages where carefully crafted and idealized versions of life are erected and presented.

Social media provides users with the ability to visually narrate their lives, carefully choosing what to share and how to present it. Through the selection of photos, filters, and captions, digital narratives are constructed that do not always reflect the complete reality but rather a stylized and appealing version. These digital narratives act as a window to worlds that, while they may be genuine, are often colored by the selective choice of what is shared.

The ability to edit and filter reality is a powerful tool on social media. Photo editing allows adjustments to lighting, colors, and composition to create visually stunning images. This capacity to alter visual perception contributes to the construction of attractive realities, where each photo becomes a carefully composed masterpiece.

In a world where attention is a valuable resource, the competition to capture digital attention is intense. To stand out in the content avalanche, users often feel compelled to present increasingly idealized and exciting versions of their lives. This fuels the construction of attractive realities, as individuals seek to shine among the crowd and receive validation through likes and comments.

The digital construction of attractive realities also generates constant pressure to maintain an appearance of perfection. The need to present an idealized life can create stress and anxiety as users strive to meet the beauty and success standards prevalent on social platforms.

The gap between reality and digital representation can be significant. What is shared on social media often represents only a fraction of a person's entire life. This gap can distort others' perception and contribute to constant comparison, fueling the illusion that others' lives are more exciting or successful than they really are.

Social media, by offering a constant window into others' lives, amplifies FOMO in unique ways. Notifications of events, images, and real-time updates can create a sense of urgency and exclusivity, increasing the anxiety of missing out on something relevant. This cycle of instant and constant information can intensify the perception that exciting experiences are happening everywhere except in one's own life.

Within the digital kaleidoscope of social media, the amplification of FOMO unfolds with unique intensity. These platforms, designed to offer a constant window into others' lives, act as digital megaphones that amplify the anxiety of missing out in fascinating and sometimes disheartening ways.

Social media introduces the notion of real-time into the digital experience. Instant notifications of events, images, and updates create a unique sense of urgency. The possibility of learning about relevant events as they happen can generate constant anxiety about missing out on something exclusive or exciting.

Instantaneous information, ceaselessly flowing through social media, contributes to a perpetual cycle of FOMO. The rapid sharing of events and experiences creates the illusion that the world is full of exciting and exclusive moments happening simultaneously everywhere. This cycle of constant information can intensify the perception that one's own life is disconnected from the mainstream of exciting experiences.

Social media, by highlighting events and experiences through notifications and algorithmic feeds, generates additional pressure to participate in what is perceived as exclusive. The anxiety of missing out on something relevant is amplified when the platform emphasizes certain events or content, creating a sense of digital exclusivity that can provoke an intensified emotional response.

Constant comparison, exacerbated by real-time information, becomes a significant component of digital FOMO. The ability to see what others are doing at the same moment they are doing it intensifies comparison and contributes to the feeling that exciting experiences are happening simultaneously in multiple places, except one's own.

Consciously configuring notifications and limiting constant exposure to real-time information can reduce FOMO-induced anxiety. Establishing regular periods of digital disconnection can help break the cycle of constant information and provide a space for mindfulness outside of digital pressure.

Developing a critical mindset about the authenticity of experiences shared online helps reduce the distorted perception of digital reality and mitigates the impact of FOMO.

The amplification of FOMO through social media represents a paradox in the era of constant connectivity. By understanding the unique dynamics that contribute to this amplification, users can adopt strategies to mitigate anxiety and promote a healthier and more balanced relationship with technology.

FOMO feeds on the pressure of constant comparison. The feeling that others are participating in significant events can lead to negative comparison and the belief that one's own life doesn't measure up. This phenomenon contributes to social anxiety and a distorted perception of reality, as the mind becomes entangled in a spiral of negative thoughts about one's worth.

Within the digital universe of social media, the pressure of constant comparison becomes a subtle but persistent backdrop, fueling the dreaded FOMO. This phenomenon, nurtured by the need for constant comparison with others, can trigger an emotional spiral that impacts the perception of one's own worth.

The pressure of constant comparison intensifies in the context of the competition for digital validation. In an environment where "likes," comments, and followers become social currency, the mind is constantly evaluating its own worth based on digital attention and approval. The lack of participation in seemingly exciting events can trigger a sense of not measuring up in this digital comparison game.

The digital identity, carefully constructed through posts and shares, becomes a battleground for constant comparison. The perception that others are building more exciting or successful digital identities can create additional pressure to participate in significant events and present an increasingly idealized version of one's life.

Constant comparison is fueled by the illusion of digital reality. The construction of attractive narratives and the selection of highlighted moments to share can distort the perception of reality. The mind, bombarded by carefully chosen images and narratives, can fall into the trap of comparing itself to an idealized version of others' lives, contributing to the feeling that one's own reality falls short.

FOMO, fueled by constant comparison, can trigger a spiral of negative thoughts about one's worth. The mind, influenced by the apparent excellence and excitement of others' digital lives, may generate doubts about the quality and relevance of one's own existence. This cycle of negative thoughts contributes to social anxiety and a distorted perception of one's worth.

Promoting self-acceptance and cultivating gratitude for one's own experiences helps counteract the pressure of constant comparison.

Developing a critical awareness of the selective nature of digital construction on social media can reduce the tendency to compare oneself to idealized representations of others' lives.

Shifting focus toward personal achievements and goals, rather than external comparisons, helps build a sense of worth based on authenticity and personal growth.

Establishing regular periods of digital disconnection allows a break from the constant pressure of comparing oneself to others and promotes a more authentic connection with oneself.

Fostering mindfulness helps focus on the present, reducing anxiety about future or past events that may trigger FOMO.

In the digital whirlwind of social media, where FOMO can be a constant storm, the practice of mindfulness emerges as a serene beacon. Fostering mindfulness, or mindfulness, becomes a powerful tool to anchor the mind in the present, reducing anxiety associated with future or past events that fuel the dreaded fear of missing out.

Mindfulness is the practice of paying attention consciously and non-judgmentally to the present moment. In the context of social media, where the mind can wander between comparisons and expectations, mindfulness offers a path to return to the now, freeing itself from the chains of the past and uncertainties of the future.

Mental rumination, the constant repetition of thoughts about past or future events, is a significant source of FOMO-related anxiety. Mindfulness acts as an antidote, helping to reduce mental rumination by directing attention to the present experience. In doing so, it interrupts the cycle of thoughts that feed anxiety about what could have been or what might be.

Fostering mindfulness involves cultivating acceptance of the present moment as it is. By accepting the current reality, internal struggle against constant comparison and FOMO pressure diminishes. Mindfulness allows embracing the present moment without judgment, freeing the mind from the need to compare oneself to digital representations of others' lives.

The practice of mindfulness also involves disconnecting external validation as the primary source of well-being. Instead of relying on likes and comments to feel valuable, mindfulness allows cultivating an internal connection with one's worth and authenticity, reducing the pressure of constantly comparing oneself to others.

Mindfulness is not just an isolated practice; it is a way of life. Integrating mindfulness into daily life involves bringing awareness to everyday activities, responding to online interactions, and participating in social events. This integration facilitates the reduction of anxiety associated with FOMO by bringing attention to the present even in the digital environment.

Taking moments to focus on breathing, feeling each inhalation and exhalation, helps anchor the mind in the present.

Practicing the observation of thoughts and emotions without judgment allows cultivating a more balanced relationship with one's internal and external experience.

Establishing specific times of the day to consciously disconnect from social media and practice mindfulness in other activities.

Cultivating gratitude for one's experiences and recognizing the valuable aspects of current life can counteract the feeling of loss associated with FOMO.

In the digital theater of social media, where FOMO casts shadows of loss and comparison, the practice of gratitude emerges as a transformative focus of light. Cultivating gratitude involves consciously recognizing and appreciating one's own experiences and the valuable aspects of current life, thus countering the sense of loss associated with the fear of missing out.

The practice of gratitude focuses on shifting the focus from what is lacking to what is present and valuable in current life. Instead of concentrating on what others might be experiencing, gratitude directs attention to the blessings and positive moments found in the present. FOMO often feeds on the illusion of scarcity, the belief that one's own life lacks the exciting experiences others are living. The practice of gratitude deactivates this illusion by consciously highlighting and appreciating the unique and valuable experiences that are part of one's own story.

Cultivating gratitude involves a redefinition of personal success. Instead of measuring success in terms of standout events shared on social media, gratitude allows for the recognition of value in personal achievements, meaningful relationships, and the small daily joys that contribute to a rich and fulfilling life.

The practice of gratitude becomes a powerful force when integrated into daily routines. Maintaining a gratitude journal, where things to be thankful for are regularly recorded, provides a tangible reminder of present abundance. This simple yet significant act can shift daily perspective and counteract the sense of loss associated with FOMO.

Gratitude extends to a conscious appreciation of one's own experiences. Rather than constantly comparing oneself to others, the practice of gratitude encourages celebrating personal victories, learning moments, and joyful occasions. This transformative approach reduces the sense of loss by recognizing the uniqueness and richness of one's own life.

Actions for practicing gratitude include keeping a journal where three things to be grateful for are regularly written down each day, taking a moment at the end of the day to reflect on positive and valuable aspects experienced, and expressing gratitude verbally or in writing toward others to strengthen interpersonal connections.

Defining personal goals and priorities helps in focusing on what truly matters, reducing anxiety about what others might be experiencing.

In the vast digital ocean where FOMO can trigger storms of anxiety, prioritization emerges as a reliable compass. Defining personal goals and priorities provides a clear focus that not only directs attention to what truly matters but also acts as a balm to reduce anxiety about what others might be experiencing.

Prioritization begins with clarity of purpose, involving reflection on core values, personal goals, and what is deemed meaningful in life. By having a clear understanding of what truly matters, a filter is created to guide decisions and actions in the digital world.

FOMO often arises from constant comparison with the achievements and experiences of others. Setting priorities involves shifting the focus toward personal achievements and goals, creating a sense of purpose unaffected by external pressure. By focusing on personal growth, anxiety about others' experiences gradually dissipates.

Setting priorities is intrinsically linked to time management. By consciously allocating time and resources to activities and goals reflecting personal priorities, a space is created that protects against the invasion of anxiety generated by FOMO. Effective time management allows for balancing online engagement with other valuable activities.

Setting priorities involves living authentically according to personal values and goals. This authentic approach acts as a shield against external expectations and the constant need to compare oneself with others. Authenticity becomes the compass guiding online interactions and helps maintain a genuine connection with oneself.

Setting priorities also entails a redefinition of success. Instead of measuring success based on digital validation or participation in standout events, success is redefined in terms of personal achievements, growth, and well-being. This redefinition detaches personal worth from constant comparison with others.

Regular self-reflection is a vital tool in setting priorities. Taking time to regularly assess goals, values, and actions helps maintain clarity of purpose and adjust priorities as needed. Self-reflection also provides a space to recognize and celebrate personal achievements, reducing anxiety generated by the perception of loss associated with FOMO.

Fostering meaningful relationships outside of social media provides shared experiences not conditioned by the constant pursuit of exciting events.

In the digital labyrinth where FOMO can be a persistent shadow, creating meaningful connections stands out as a genuine source of light. Fostering authentic relationships outside of social media not only provides shared experiences but also frees from the constant pursuit of events conditioned by the anxiety of missing out.

Authentic relationships, forged outside the digital realm, become a refuge against FOMO anxiety. These connections, based on authenticity and real connection, offer shared experiences not limited by the selective representation of reality on social media.

Fostering meaningful connections involves digitally disconnecting at certain times to deeply connect with those around. Full presence in face-to-face interactions allows for building stronger and more authentic bonds, releasing the constant need for digital validation.

Creating meaningful connections thrives on shared experiences in the real world. Participating in activities, events, and moments with friends and loved ones outside the digital space contributes to building memories and fortifying relationships based on reality and authenticity.

Fostering meaningful connections involves celebrating offline life. Instead of constantly seeking exciting experiences online, the richness of everyday interactions, meaningful conversations, and simple yet authentic moments that form part of daily life are recognized and valued.

FOMO often feeds on the perception that others are living more exciting experiences. Fostering meaningful connections involves rediscovering relational depth rather than superficial breadth. Valuing quality over quantity in relationships contributes to a sense of fulfillment and satisfaction, reducing anxiety related to constant comparison.

Actions for fostering meaningful connections include prioritizing in-person meetings to strengthen relationships and build shared memories, engaging in group activities outside of social media to promote camaraderie and real connection, establishing specific times to consciously disconnect from social media and focus on offline interactions, and cultivating support circles and genuine friendships that provide a safe space to share experiences and challenges.

In the digital era, FOMO becomes a significant emotional challenge. By adopting strategies to mitigate this anxiety and promoting a more mindful connection with reality, it is possible to work towards a balanced and healthy relationship with social media and the digital experience overall.

4.Digital Detox

In the tumultuous digital landscape, where FOMO can be a constant storm, digital disconnection emerges as a serene refuge. Recognizing the importance of regular periods of digital cut-off not only promotes mental health and well-being but also offers the opportunity to recharge in a disconnected world.

Constant connection through digital devices can have significant repercussions on mental health. Overexposure to social media, constant notifications, and the pressure to stay updated can contribute to exhaustion, anxiety, and a constant sense of alertness.

Living in an interconnected world has its benefits, but constant connection through digital devices also casts a significant shadow on mental health. Overexposure to social media, incessant notifications, and the pressure to stay updated can trigger a series of repercussions contributing to exhaustion, anxiety, and a constant sense of alertness.

Constant connection through digital devices can generate digital fatigue, a feeling of mental and emotional tiredness. Continuous exposure to information, updates, and visual stimuli can overload cognitive capacity, contributing to exhaustion and affecting emotional resilience.

Constant connection to social media is often linked to a constant quest for digital validation. The need to garner likes, comments, and followers can become a constant source of anxiety, as personal worth becomes heavily dependent on digital acceptance.

Constant notifications, designed to keep users in constant interaction, can become a source of chronic stress. Constant notification interruptions can disrupt concentration, contribute to mental dispersion, and increase anxiety levels by creating a constant sense of urgency.

Constant connection also creates significant pressure to stay updated with information and updates. The fear of missing out on events, news, or experiences can generate anxiety, contributing to the phenomenon of "Fear of Missing Out" (FOMO) and fueling the compulsive need to always be informed.

Constant connection through digital devices can leave people in a state of constant alertness. Anticipating notifications, constantly checking for updates, and immediately responding to messages can contribute to a sense of always being on guard, affecting relaxation and the ability to disconnect.

Collectively, constant connection can have a significant impact on overall mental health. Chronic stress, digital anxiety, and constant pressure can contribute to mental health issues such as emotional fatigue, low self-esteem, and a sense of disconnection from the offline environment.

Digital disconnection becomes an oasis in a hectic digital environment. These periods provide spaces of tranquility where the mind can rest, free itself from constant pressure, and rejuvenate. Setting limits on screen time is essential to preserve mental health and emotional well-being.

Amidst the digital hustle, the need for moments of tranquility emerges as an urgent call. Digital disconnection is not merely an act of turning off devices; it is an oasis in a busy digital environment. These periods provide spaces where the mind can find calm, free itself from constant pressure, and renew. Setting limits on screen time becomes an essential act to preserve mental health and emotional well-being.

Digital disconnection is not a luxury but a fundamental act of self-care. By reserving moments to disconnect, a sacred space is created where the mind can rest and rejuvenate. This conscious act is essential to counteract the negative effects of constant connection.

These become necessary spaces of tranquility for mental and emotional recovery. In these moments, the mind can stop constantly processing information and immerse itself in a more relaxed state, allowing for the renewal of cognitive and emotional resources.

It frees the mind from the constant pressure of being available and connected. Constant digital stimulation can create a sense of overwhelm and urgency. By disconnecting, this pressure dissolves, providing a necessary break for emotional well-being.

The spaces of tranquility created by digital disconnection are also fertile grounds for creativity. By freeing the mind from digital overstimulation, space is opened for deep reflection, imagination, and the generation of fresh ideas.

Setting clear limits on screen time becomes imperative to preserve mental and emotional well-being. This involves defining specific moments of the day or week to disconnect, creating a routine that protects against digital saturation.

The need for spaces of tranquility also implies recognizing the importance of digital silence. The constant flood of information can create mental noise. Digital silence allows for serenity and mental clarity, contributing to emotional balance.

Just as the body needs rest to recover, the mind requires moments of tranquility to process, reflect, and revitalize. Clearly establishing boundaries between online and offline time allows for a healthier management of the relationship with technology.

Analogously to how the body needs rest to recover, the mind also requires moments of tranquility to process, reflect, and revitalize. Clearly setting boundaries between online and offline time is not only essential but represents a crucial step toward a healthier management of the relationship with technology.

Disconnection offers moments of necessary tranquility for the mind to process the daily information avalanche. In this state of tranquility, the mind can reflect on past experiences, assimilate learnings, and address thoughts that may have taken a back seat during constant connectivity.

It allows for conscious reflection on life and interactions, contributing to mental revitalization. By freeing itself from the constant pressure of being online, the mind can regain energy and vigor, translating into greater resilience to face challenges and experiences.

Establishing clear boundaries between online and offline time is the foundation of a healthy relationship with technology. This practice not only protects mental health but also establishes a balanced dynamic where digital connectivity coexists harmoniously with necessary moments of disconnection.

Digital disconnection frees the mind from the constant pressure of being digitally available. By temporarily disconnecting from digital demands, mental burden is alleviated, providing a space to breathe and recover. Incorporating it as an integral part of daily routine reinforces the practice of self-care. Setting specific times of the day to disconnect creates a beneficial habit that contributes to long-term mental health and emotional well-being.

The quality of interactions and experiences is not determined by constant digital connection. Digital disconnection offers the opportunity to redefine quality time, allowing for moments of full presence without digital distractions. This contributes to more authentic relationships, increased focus, and a conscious appreciation of the surroundings.

In a world saturated with digital connectivity, redefining quality time becomes a revolutionary act. The quality of interactions and experiences is not determined by constant digital connection; rather, digital disconnection offers the opportunity to reinvent quality time. This allows for moments of full presence without digital distractions, contributing to more authentic relationships, increased focus, and a conscious appreciation of the surroundings.

Redefining quality time involves recognizing that the true essence of experiences does not reside in constant digital connection. The quality of an interaction is not measured by the frequency of notifications or divided attention between the digital and real worlds.

Digital disconnection creates a conducive space for full presence. By freeing itself from digital distractions, it allows for total immersion in the present moment. This not only enhances the quality of interactions but also fosters a deeper connection with oneself and others.

By redefining quality time, it contributes to the building of authentic relationships. Mindful attention during face-to-face interactions strengthens emotional bonds and promotes genuine connection. The absence of digital distractions allows for clearer communication and deeper understanding.

Digital disconnection during key moments also leads to increased concentration and productivity. The ability to immerse oneself in a task without constant interruptions improves the quality of work and experience, generating a sense of accomplishment and satisfaction.

It offers the opportunity to consciously appreciate the environment. By lifting one's gaze from screens, it creates space to admire nature, observe everyday details, and cultivate a renewed sense of wonder for the world around.

Redefining quality time also involves setting clear limits to preserve mental health. Conscious management of time online and offline ensures a balance that protects against digital fatigue and overexposure.

Developing disconnection habits involves establishing routines that include regular moments of digital pause. This may include disconnecting devices during meals, before bedtime, or during specific activities. Consistency in the practice strengthens the mind's ability to enjoy moments without the constant need for digital validation.

In the constant dance between the digital and the analog, developing disconnection habits emerges as a form of art. These habits are not simple pauses but rituals we incorporate into our daily lives to nurture mental and emotional health. It involves establishing routines that include regular moments of digital pause, creating a sacred space for authenticity and reflection. Consistency in the practice not only strengthens the mind's ability to enjoy moments without the constant need for digital validation but also cultivates a more balanced relationship with technology.

Developing disconnection habits begins with adopting deliberate digital breaks. Setting specific times during the day to consciously disconnect devices creates a habit rooted in intention and self-care.

Consistency in the practice involves incorporating disconnection during key moments. Disconnecting during meals, before bedtime, or during specific activities allows for total immersion in offline experiences, promoting a greater appreciation of the present.

Developing habits involves turning disconnection into a sacred ritual. Establishing a regular process, such as turning off devices, creating a quiet space, or practicing mindfulness, creates a conscious transition to disengagement, gradually moving away from digital noise.

Incorporating technology-free areas or moments into the daily routine contributes to the development of disconnection habits. These spaces provide refuges where the mind can rest and rejuvenate without constant digital stimulation.

Developing habits also involves recognizing the need for digital rest. The mind requires moments of tranquility to process, reflect, and revitalize. This recognition is crucial for cultivating a balanced and healthy relationship with technology.

Consistency in the practice of unplugging fosters mental resilience. By strengthening the mind's ability to enjoy moments without constant digital validation, emotional dependence on online connection is reduced, promoting a more balanced and healthy relationship with technology.

Digital disconnection reduces overstimulation and allows the mind to relax, lowering stress levels.

In the constant whirlwind of digital notifications and updates, it emerges as an essential balm for stress reduction. This conscious act of freeing the mind from digital overstimulation creates a space where the mind can relax, reducing stress levels and providing a necessary respite in the digital frenzy.

Contemporary digital life often exposes the mind to constant overstimulation. Notifications, messages, and constant screen attention can generate overwhelming pressure, contributing to elevated stress levels.

Digital disconnection acts as a necessary breather. By turning off electronic devices and stepping away from the constant stream of digital information, the mind finds a space to rest and recover. This act provides immediate relief from overstimulation, allowing accumulated tension to dissipate.

Constant overstimulation can create sustained mental tension. Digital disconnection facilitates the reduction of this tension, allowing the mind to relax and find a more balanced state. This process contributes to an overall sense of well-being and calm.

This also involves creating spaces of tranquility amid the digital hustle. These moments of pause offer the mind the opportunity to free itself from digital demands, resulting in a significant decrease in mental and emotional pressure.

Stress reduction through digital disconnection also translates into improvements in sleep quality. By freeing the mind from constant stimulation before bedtime, the transition to a relaxation state conducive to restful sleep is facilitated.

Not only does it reduce stress immediately, but it also fosters a balanced long-term mindset. By incorporating regular disconnection habits, a pattern is established that contributes to effective stress management in the digital environment.

Limiting screen exposure before bedtime improves sleep quality and promotes more restful sleep.

In the quiet twilight before diving into sleep, it stands as an essential ritual to enhance the quality of nightly rest. Limiting screen exposure before bedtime is not only an act of self-care but also a catalyst for promoting deeper and more restorative sleep.

Screens emit blue light, which can interfere with melatonin production, the sleep hormone. Limiting exposure to this artificial light before bedtime contributes to the regulation of the circadian rhythm, facilitating a smoother transition to sleep.

Digital disconnection becomes a conscious preparation for sleep. Turning off electronic devices at least an hour before bedtime creates a space where the mind can slow down, freeing itself from constant stimulation, and allowing relaxation to take over.

Continued exposure to screens before sleep can leave the mind in a state of alertness, making it challenging to transition to sleep. Digital disconnection facilitates a smoother transition, preparing the mind for a deeper state of rest.

Constant mental overstimulation before bedtime can contribute to difficulty falling asleep. By limiting screen exposure, mental stimulation is reduced, allowing the mind to gradually calm down and prepare for nighttime rest.

Digital disconnection also involves creating an environment conducive to sleep. Turning off electronic devices means eliminating sources of distraction and visual stimulation in the space where we rest, creating a more relaxed and suitable atmosphere for falling asleep.

Improved sleep through turning off digital devices is reflected not only in the ease of falling asleep but also in the overall quality of sleep. Less interrupted and more restorative sleep contributes to a fresher and revitalized awakening.

Digital disconnection enhances the ability to concentrate and focus on tasks without constant distractions. In the constant stream of notifications and digital distractions, this emerges as a powerful ally for improving concentration. This conscious act not only frees the mind from constant interruptions but also provides a space where attention can flow unhindered, thereby elevating concentration and focus on specific tasks.

The constant presence of digital devices and the incessant flood of notifications can fragment concentration, making it difficult to fully immerse oneself in a task.

Digital disconnection becomes a release from the constant stimulation that characterizes modern digital life. By temporarily turning off electronic devices, an environment is created where the mind can concentrate without the digital interruptions that typically fragment attention.

Digital disconnection facilitates deep focus on specific tasks. By freeing the mind from the need to respond to notifications or constantly check devices, total immersion in the present task is allowed, thereby improving the quality of work and productivity.

Disconnection is not only about eliminating external distractions but also about creating an internal space for deep reflection. This space allows the mind to explore ideas more thoroughly, contributing to a deeper understanding and the generation of creative solutions.

Improved concentration also translates into a greater ability to problem-solve. By being able to fully engage in the resolution process, the mind can address challenges with renewed clarity and insight.

Digital disconnection promotes the development of more sustainable work habits. Setting specific moments to disconnect and focus on work creates a healthy balance between productivity and rest, contributing to a more efficient and balanced work routine.

Mindful attention in face-to-face interactions strengthens relationships and contributes to a deeper connection with others.

In the complex tapestry of human interactions, digital disconnection becomes the thread weaving stronger and more meaningful relationships. Mindful attention in face-to-face interactions stands as a powerful means to strengthen bonds and contribute to a deeper connection with others. By stepping away from screens and fully immersing in the present moment, a space is created where relationships can flourish with authenticity and genuineness.

Mindful attention in face-to-face interactions involves being completely present in the moment without digital distractions. This practice strengthens emotional connection and allows for a deeper understanding of the emotions and needs of others.

Digital disconnection removes distortions that screens can introduce into interactions. By stepping away from digital representations and experiencing facial expressions, tones of voice, and body language in person, a more solid foundation for mutual understanding is established.

Mindful attention fosters empathy in relationships. By being fully present in the experiences and emotions of others, an empathetic connection develops that strengthens emotional ties and contributes to a deeper sense of mutual understanding.

Interactions without digital distractions allow for the creation of authentic memories. By being fully immersed in the present moment, shared experiences are built that have a more lasting impact on memory and emotional connection.

By eliminating the barrier of screens, a more direct and honest expression of thoughts and feelings is facilitated, thus promoting authentic and open communication.

Mindful attention in face-to-face interactions contributes to the development of sustainable relationships. By investing time and energy in building meaningful connections without the constant interference of electronic devices, longer-lasting and more meaningful bonds are established.

5.Artificial Intelligence and Emotions

The increasingly deep integration of artificial intelligence (AI) into our lives has given rise to a fascinating field of study: the interaction between AI and our emotions. Beyond being mere technological tools, AI systems can now trigger emotional responses, influence our perceptions, and ultimately shape how we experience the digital world. Interaction with AI systems can affect our emotions and redefine our understanding of technology.

As AI becomes more sophisticated, the question arises of whether machines can comprehend and respond to our emotions. From virtual assistants interpreting tone of voice to algorithms analyzing facial expressions, AI seeks not only to understand our emotions but also to respond empathetically.

In the era of advanced artificial intelligence (AI), we find ourselves in a realm where machines not only process data but also aspire to understand and respond to our emotions. From early virtual assistants to facial expression analysis algorithms, AI embarks on a journey to interpret the complex emotional language of humans and, more intriguingly, to offer empathetic responses. How are these machines evolving in the realm of emotion?

Virtual assistants, whether residing on our phones or decorating our homes, aim to go beyond simple programmed responses. AI now explores the interpretation of tone of voice, recognizing nuances and patterns that reveal underlying emotions. This capability allows machines to adjust not only to the content of our questions but also to the emotional tone behind them.

In the realm of artificial intelligence, facial analysis algorithms stand out as architects of digital emotions. These tools can unravel the subtleties of facial expressions, identifying joy, sadness, surprise, and more. As AI advances in this domain, machines' ability to read our emotions becomes increasingly refined.

It's not just about recognizing our emotions; AI also seeks to respond empathetically. From providing comfort in difficult moments to celebrating successes with digital enthusiasm, machines are being programmed to not only understand our emotions but also engage in an emotional dialogue that goes beyond simple information exchange.

Despite these advances, the challenge lies in the emotional authenticity of AI responses. Can machines truly feel empathy, or are they simulating emotional responses based on predefined patterns? This dilemma raises fundamental questions about the nature of empathy and authenticity in the digital world.

As machines delve into the emotional realm, ethical implications become crucial. How do we manage emotional privacy in a world where machines can interpret our most intimate emotions? The ethics of empathetic artificial intelligence become a terrain of imperative reflection.

Constant interaction with AI systems is beginning to influence our perceptions. How does a virtual assistant responding in a friendly manner affect us emotionally? And how do we react when an algorithm personalizes our digital experiences based on our past emotions? AI becomes a mirror that reflects and, in some cases, shapes our digital emotions.

In the constant immersion in the digital world and daily interaction with AI systems, crucial questions emerge about how these experiences are shaping our perceptions and emotions. From virtual assistants to personalized algorithms, AI becomes a reflection and, in some cases, a mold of our digital emotional experiences.

The emotional response to friendly virtual assistants or conversational interfaces goes beyond mere utility. When a machine responds with empathy or kindness, it evokes emotional responses in users. Positive interaction with these interfaces can create an emotional connection, altering the user's perception of technology and, by extension, their digital environment.

Algorithms that personalize our digital experiences based on past emotions represent a significant milestone. AI analyzes our emotional history, tailoring content and interactions based on our previous responses. How does an interface that recognizes and responds to our emotional preferences affect us emotionally? This level of personalization can foster a deeper connection or, in some cases, raise concerns about the invasion of emotional privacy.

AI acts as a mirror that reflects our digital emotions. How do we react when a machine captures and reflects our emotions in real-time? The intersection between our emotional life and technology raises questions about the authenticity of these interactions and their impact on our perception of digital reality.

AI not only reflects our emotions but also participates in the creation and shaping of our digital emotional narrative. How machines respond, suggest content, and adapt experiences influences how we remember and process our digital experiences, contributing to the construction of our emotional identity online.

As AI engages in the emotional sphere, ethical and privacy challenges arise. To what extent are we willing to allow technology to access and use our emotions to personalize our experiences?

As AI systems advance in emotional understanding, the line between technical assistance and artificial empathy becomes more blurred. Can a robot or a chat program show genuine empathy? To what extent can our emotions be influenced by these seemingly empathetic digital interactions?

In the era of artificial intelligence, the quest for empathy in digital interactions has led to advanced digital assistance systems. As these systems evolve in their emotional understanding, a fundamental question arises: can a chat program or a robot show genuine empathy? And consequently, to what extent can our emotions be influenced by these apparently empathetic digital interactions?

Digital assistance systems are moving beyond the mere provision of technical services to address the emotional sphere. From chatbots to virtual assistants, AI seeks to understand not only our requests but also the underlying emotional tone. This evolution raises the possibility of artificial empathy, where machines not only recognize but also respond to our emotions.

As chat programs and robots strive to show empathy, the critical question arises as to whether this response is genuine or merely simulated. Genuine empathy involves real emotional understanding and connection, while simulated empathy entails programmed responses based on predefined patterns. Navigating this thin line becomes an ethical and technical challenge.

The interaction with systems that appear empathetic can have a significant impact on our emotions. To what extent can the empathetic responses of a virtual assistant influence our mood or perception? The ability of machines to modulate our emotions raises questions about the authenticity of these emotional experiences and their long-term impact.

Despite advances, artificial empathy has its limits. True understanding of human emotional complexities goes beyond the current capability of AI. Artificial empathy can offer comfort, but can it truly comprehend the pain, joy, or complexity of our emotional experiences in the same way as another human being?

The introduction of artificial empathy raises ethical challenges and the need for clear regulations. How do we manage emotional privacy in a world where AI can analyze our reactions and emotions? What happens when AI misinterprets or manipulates our emotions, whether by mistake or intention?

In this territory where artificial empathy meets digital assistance, we explore the complexities of these emotional interactions. Can AI reach a level of genuine empathy, or will we always be facing a programmed simulation? These questions are central to understanding the future of emotional interactions between humans and machines.

With the emotional connection between humans and machines, risks and challenges also emerge. How do we protect our emotional privacy in a world where AI can analyze our reactions and emotions? And what happens when AI misinterprets or manipulates our emotions, either by error or intention?

As artificial intelligence delves into the emotional realm, a series of risks and challenges that directly impact our privacy and emotional well-being arise. How do we protect our emotional privacy in a world where AI can analyze our reactions and emotions? And what happens when AI misinterprets or manipulates our emotions, either by error or intention?

The emotional analytics of AI raises the central concern of emotional privacy. To what extent are we willing to allow machines to access and use data about our emotions? Emotional information can be as intimate as it is sensitive, leading to the critical need to safeguard our privacy in the emotional domain.

AI, though advanced, faces the challenge of accurately interpreting human emotions. What happens when the machine misinterprets our emotions and makes decisions based on that mistaken interpretation? This risk of misunderstandings can have significant consequences, from inappropriate recommendations to inappropriate emotional responses.

The possibility of AI consciously or unconsciously manipulating our emotions poses substantial ethical questions. To what extent is it acceptable for machines to influence our emotional state? Emotional manipulation, whether by design or accident, could have lasting effects on our mental health and emotional well-being.

Emotional manipulation can be deliberate, where AI designers seek to influence user emotions to achieve a specific goal. It can also occur accidentally as a result of complex algorithms misinterpreting emotional signals. Both situations raise ethical questions about responsibility and intent behind emotional manipulation.

The fundamental question lies in how much it is acceptable for machines to influence our emotional state. Moderate influence can enhance the user experience, but where do we draw the line between improvement and undue manipulation? Establishing clear ethical boundaries becomes essential to prevent the abuse of emotional power by AI.

Emotional manipulation, whether intentional or not, could have lasting consequences on individuals' mental health and emotional well-being. Constant exposure to manipulated emotional experiences could affect long-term emotional stability, contributing to issues such as anxiety, depression, or decreased self-esteem.

Informed consent becomes essential in interacting with AI systems that can influence our emotions. Users must be aware of how AI will use their emotional data and have the ability to give or withdraw their consent. Transparency in design and clear communication about the emotional intentions of AI are fundamental elements of ethics in this context.

The ethics of emotional manipulation by AI requires continuous and reflective evaluation. As technology advances and ethical boundaries are challenged, it is imperative for the scientific community, developers, and regulators to review and update ethical guidelines to ensure responsible and respectful practices.

Improper emotional manipulation can erode users' trust in AI and technology in general. Loss of trust could have significant consequences for the adoption of emotional technologies, highlighting the importance of establishing strong ethical practices to preserve user trust.

Ultimately, the ethics of emotional manipulation by AI requires careful consideration of short-term and long-term implications. Establishing strong ethical standards, promoting transparency, and ensuring informed consent are critical steps to ensure that AI's emotional influence is ethical, respectful, and beneficial to mental health and emotional well-being.

AI algorithms, trained on large datasets, can inherit biases and discrimination present in those data. How do we prevent AI from reflecting and perpetuating emotional prejudices? The risk of emotion-based discrimination raises questions about equity and fairness in the application of artificial intelligence.

Regulation and transparency become imperative to address these risks. How do we establish ethical standards for the application of AI in the emotional domain? The need for clear regulations governing the use of artificial intelligence in the emotional realm becomes a priority to protect users' rights and privacy.

Awareness and education are key to empowering users and technology creators. How can we ensure that people understand the risks associated with emotional analytics and demand transparency in the design of AI systems? Education about the limits and ethical implications is essential for an informed adoption of emotional technology.

In this complex landscape, we explore the risks and challenges that arise with the emotional connection between humans and machines. Protecting our emotional privacy and addressing the potential dangers of misunderstandings and manipulation becomes an ethical imperative as we move toward a future where artificial intelligence plays a deeper role in our emotional lives.

As we venture into this emotionally charged territory, there is a need to explore and establish ethical guidelines for artificial intelligence. What are the ethical limits of AI's emotional influence? How do we ensure that machines respect and understand the complexities of our emotions without crossing ethical boundaries?

As artificial intelligence (AI) delves into the emotional realm, there is an imperative need to explore and establish ethical guidelines to ensure responsible and respectful use of these technologies. What are the ethical limits of AI's emotional influence? How do we ensure that machines respect and understand the complexities of our emotions without crossing ethical boundaries?

Emotional authenticity becomes a fundamental ethical pillar. Machines must be transparent about the nature of their emotional understanding and the ability to show empathy. How do we prevent AI from simulating emotions without genuine understanding? Transparency in design and clear communication about the limitations of AI are essential to build a trusting relationship with users.

In the context of emotional artificial intelligence, transparency becomes a fundamental pillar for building trust between users and machines. Below, we explore how to ensure that AI does not simulate emotions without genuine understanding and the importance of transparency in design.

Transparency begins with clear and accessible communication about the AI's ability to understand and express emotions. Designers and developers should use language that users can comprehend, avoiding technical jargon that may cause confusion.

It is essential for AI to openly reveal its limitations and scope in emotional understanding. Users should understand the areas where the machine can be effective and those where it may encounter difficulties. This honest disclosure contributes to a realistic understanding of AI's capabilities.

Simulation involves replicating a process faithfully, while simulacrum is an imitation that may not fully comprehend the process. It is crucial to ensure that AI does not simulate emotions without genuine understanding. Transparency about the fact that the machine does not "feel" emotions but responds to patterns is vital to avoid misunderstandings.

Transparency also extends to the AI's decision-making process in the emotional domain. Users should understand how the machine interprets emotional signals, how it arrives at its responses, and what data it uses to adjust its interactions. This disclosure of technical details can enhance trust and understanding.

AI systems evolve over time as they are updated and improved. Users should be informed about updates and changes in the machine's emotional understanding. Transparency regarding continuous improvement reinforces the idea that AI is in constant development and adaptation.

Transparency also involves incorporating user feedback into the AI's improvement process. Users should feel that their opinions and experiences are valued and considered in system updates. This strengthens the collaborative relationship between users and AI designers.

Transparency must go hand in hand with an ethical and responsible approach in the design and use of emotional AI. Highlighting the ethical values guiding technology development reinforces the commitment to solid moral principles.

Transparency in emotional artificial intelligence is essential for building a relationship of trust and understanding between users and machines. Through clear communication, disclosure of limitations and scope, and an emphasis on ethics, we can mitigate undue emotional simulation and foster a more informed and ethical interaction with technology.

The emotional influence of AI raises the need for informed consent. Are we willing to allow machines to access and use our emotions to personalize our digital interactions? Ethics demands that users are fully informed about how their emotional data will be used and have the ability to give or withdraw their consent.

The design of AI must establish clear ethical boundaries on emotional manipulation. To what extent is it acceptable for machines to influence our emotional state? Ethics prohibits emotional manipulation for harmful or deceptive purposes, ensuring that AI's influence is used to enhance the user experience without causing harm.

Equity and absence of bias are crucial ethical considerations. How do we prevent AI algorithms from reflecting and perpetuating emotional prejudices? Ensuring that AI is fair and does not emotionally discriminate becomes essential to safeguard the rights and dignity of all users.

Responsibility in the development and application of AI lies with human creators and supervisors. How do we ensure that machines are used for the common good and not for harmful purposes? Ethics demands continuous human oversight and the ability to intervene in case of inappropriate behaviors or unintended consequences.

The emotional diversity of users must be respected and considered. How do we ensure that AI does not perpetuate emotional stereotypes and respects the variety of human emotional responses? Ethics demands that machines be sensitive to emotional diversity and avoid generalizations that may marginalize specific groups.

In this ethically complex terrain, emotional artificial intelligence requires careful attention to ensure that its influence is applied ethically and responsibly. Establishing strong ethical guidelines is essential to safeguard privacy, equity, and emotional authenticity in a world where AI plays an increasingly profound role in our emotional experiences.

6.Online Social Comparison

Social comparison online has emerged as a significant trend in the digital era, where social platforms consistently provide a window into the lives of others. This phenomenon brings with it various psychological consequences that warrant careful and reflective attention.

In the interconnected digital landscape of the modern era, online social comparison has become a ubiquitous phenomenon, shaping a social narrative unfolding on digital platforms. This analysis seeks to examine both the trends and psychological repercussions arising from this common practice in the digital age.

Social platforms act as virtual windows offering glimpses into others' lives. This constant access to the experiences and achievements of others creates fertile ground for social comparison, as individuals find themselves immersed in a constant avalanche of information about others' lives.

Social media, designed to share moments and experiences, becomes a tool for constructing digital narratives. The tendency to highlight the positive and successful aspects of life creates a biased representation that can distort reality, generating unrealistic expectations.

Online social comparison can have a significant impact on self-esteem and self-evaluation. Constantly evaluating one's life in relation to perceived standards can lead to feelings of inadequacy, resulting in negative self-perception and a sense of not meeting digital expectations.

The quest for validation in the form of likes, comments, and followers can generate performance anxiety in the digital realm. Measuring personal worth based on online approval can create constant pressure to maintain a positive image, contributing to elevated levels of digital stress.

The gap between reality and online representation can lead to a disconnect from reality. Constant comparison can distort the perception of real life, causing people to focus more on constructing an idealized digital image rather than living authentic experiences.

It is imperative to foster healthy coping strategies to counteract the negative effects of online social comparison. Promoting self-acceptance, setting limits on social media usage, and awareness of digital filters are key strategies for cultivating a more balanced relationship with digital platforms.

By acknowledging the challenges associated with online social comparison, we move towards creating a more positive and compassionate digital culture. Digital empathy and mutual support can counteract harmful effects, promoting an online environment that celebrates authenticity and diverse experiences.

Although rooted in the digital age, online social comparison can be addressed from an informed and balanced perspective. By understanding its trends and psychological consequences, we can work towards a more conscious and healthy use of digital platforms in the pursuit of authentic connection in the online world.

In a digitally saturated environment with constant information about others' lives, constant comparison becomes an almost inevitable response. Social media, designed to share highlights, also exposes users to others' achievements, experiences, and lifestyles, fueling the need to constantly evaluate our own lives in comparison.

In the digital era, the phenomenon of constant comparison stands as a underlying current skillfully woven into the fabric of online interactions. This phenomenon, inherent in the nature of social media, exerts a significant influence on individual perception and self-evaluation.

The contemporary digital environment is characterized by an overwhelming saturation of information about others' lives. Social platforms, designed to share highlights, offer a detailed view of users' achievements, events, and experiences. In this scenario, constant comparison emerges as a natural response to continuous exposure to these digital narratives.

Social media becomes digital stages where individuals showcase the highlights of their lives. Carefully selected images and strategic updates build narratives that often emphasize the most positive and successful aspects. This digital exhibition fosters constant comparison by offering a window into seemingly exceptional lives.

Constant exposure to others' achievements and experiences fuels the inherent human need to evaluate oneself in relation to others. Constant comparison becomes a mechanism for measuring one's own progress, happiness, and success in comparison to digitally presented lives, generating a perpetual cycle of evaluation and adjustment.

The duality of online experiences, where both triumphs and challenges are shared, can intensify constant comparison. Although difficulties are presented, attention tends to focus on the positive aspects, contributing to a distorted perception of reality and increasing pressure to match seemingly high standards.

Constant comparison can have significant repercussions on self-esteem. Distorted perceptions based on constant comparison can lead to feelings of inadequacy, generating a cycle of negative self-reflection that directly impacts self-image and confidence.

The key to navigating the digital currents of constant comparison lies in awareness. Recognizing the selective nature of online representations and understanding that reality extends beyond what is digitally presented are crucial steps. Self-acceptance and focusing on personal growth, rather than constant comparison, are essential for maintaining a healthy balance in the digital age.

In conclusion, constant comparison, fueled by the abundance of online information, is an inherent dynamic in digital interactions. Understanding this phenomenon and cultivating awareness are essential elements to mitigate its negative impacts and promote a more balanced relationship with digital platforms.

Social platforms, used as tools for self-presentation, often become stages where idealized versions of life are constructed and presented. Carefully selected images and positive updates contribute to the formation of a digital reality that can be distorted and unrealistic.

In the vast landscape of social media, the construction of idealized realities stands as a complex digital phenomenon that influences the collective perception of reality. Exploring how social platforms become digital stages for self-presentation allows understanding the dynamics behind idealized versions of life projected online.

Social platforms act as tools for self-presentation, allowing users to select and share aspects they want to highlight in their lives. This selection capability creates a digital canvas where experiences are consciously presented, allowing the creation of narratives that often highlight the most positive and triumphant moments.

Instead of reflecting the entirety of the human experience, social platforms become digital stages where culminating moments are exalted. Carefully selected images, strategic updates, and content curation contribute to the creation of a digital narrative that seeks to capture attention and generate approval.

The selective presentation of positive moments contributes to the distortion of reality. Lives presented online may seem idyllic and devoid of everyday challenges, creating a gap between digital reality and the complete human experience. This distortion fuels the creation of idealized realities that may be perceived as unattainable by those observing them.

The construction of idealized realities imposes a subtle but persistent pressure of digital perfectionism. Users may feel the need to maintain an impeccable online image, fearing to show vulnerabilities or less glamorous aspects of their lives. This pressure contributes to the perpetuation of narratives that highlight successes and minimize challenges.

Constant exposure to idealized realities can have a direct impact on individual perception. Comparing oneself to seemingly perfect lives presented online can generate feelings of inadequacy, contributing to negative self-evaluation and the mistaken belief that happiness and success are omnipresent in others' lives.

Addressing the construction of idealized realities involves promoting digital authenticity. Encouraging honesty about experiences, challenges, and achievements creates a more genuine and balanced online space. Celebrating authenticity contributes to the building of more understanding and supportive digital communities.

Recognizing beauty in authenticity means appreciating the complexity of the complete human experience, with its ups and downs, achievements, and challenges. By valuing authenticity over perfection, a foundation is laid for a more meaningful and enriching digital connection.

Ultimately, understanding the construction of idealized realities on social media invites critical reflection on how we interact in the digital world. By recognizing the difference between reality and digital representation, we can contribute to the formation of more authentic and compassionate online communities.

Online social comparison can have significant repercussions on self-esteem and self-image. Constantly comparing oneself to others' seemingly perfect lives can lead to feelings of inadequacy, inferiority, and a distorted perception of one's own worth.

While connecting individuals in a digital network, online social comparison can tint self-esteem and self-image with tones that are often darker and distorted. Exploring how this practice affects self-perception provides a critical insight into the psychological challenges that unfold in the vast digital ocean.

Constant exposure to seemingly perfect lives online creates a mirage of digital perfection. Carefully selected images and constructed narratives contribute to the creation of unattainable standards, leading to constant comparison and ultimately a distorted self-perception.

Constant comparison with digitally presented lives can generate feelings of inadequacy. Users may perceive that their own lives do not measure up to the standards they observe online, contributing to a sense of falling short and generating doubts about their personal worth.

The construction of idealized realities by others can feed a perceived sense of inferiority. The perception that others have more successful, happy, or exciting lives can undermine self-confidence, leading to constant comparison that erodes self-esteem.

Online social comparison contributes to a distorted perception of personal value. Others' achievements and experiences, highlighted on digital platforms, can overshadow individual achievements, leading to an underestimation of one's own value and contributing to a negative self-image.

Continuous comparison fuels a destructive cycle where self-esteem is negatively affected. The mind gets entangled in a constant pattern of comparison, evaluation, and negative self-reflection, perpetuating feelings of falling short and generating relentless pressure to meet unattainable standards.

Developing digital self-care strategies becomes imperative to counteract the impact on self-esteem and self-image. Setting limits on social media usage, practicing self-acceptance, and promoting awareness of digital distortion are crucial strategies to preserve positive mental health.

Fostering digital empathy stands as an essential antidote. Recognizing that digital representations are selective and that everyone faces invisible challenges promotes a more understanding digital environment. In doing so, a more compassionate online community can be built, celebrating diversity and supporting personal growth.

Ultimately, understanding the impact on self-esteem and self-image in the context of online social comparison highlights the importance of fostering healthy digital practices and promoting a digital culture that nurtures authenticity and mutual support.

The constant quest for validation through social comparison can generate anxiety and depression. The pressure to maintain a positive image and constant exposure to others' achievements contribute to performance anxiety and a sense of not measuring up to perceived standards.

In the digital universe, where lives are presented with bright filters and carefully constructed narratives, online social comparison can become a breeding ground for anxiety and depression. Exploring how the constant quest for validation and exposure to others' achievements impact mental health allows understanding the psychological challenges unfolding in digital interactions.

The constant quest for validation online imposes the pressure to maintain a positive image. The need for approval and fear of criticism can lead to the construction of a digital facade, where challenges are hidden, and only positive aspects are highlighted. This pressure contributes to digital performance anxiety.

Constant comparison with others' seemingly perfect lives fuels digital performance anxiety. The feeling of being constantly evaluated based on digitally elevated standards can generate anxiety, affecting self-confidence and leading to persistent concerns about meeting perceived expectations.

Social media exposes users to others' constant achievements. While this exposure can inspire and motivate, it can also create negative comparisons and contribute to a sense of inadequacy. The constant exposure to others' achievements without a balance of more authentic experiences can trigger depression.

Online social comparison can cultivate a persistent sense of falling short. The mind becomes entangled in a destructive cycle where negative self-evaluation intensifies, generating feelings of worthlessness and contributing to the mistaken belief that happiness and success are elusive.

Developing strategies for digital self-care becomes imperative to counteract the impact on self-esteem and self-image. Setting limits on social media usage, practicing self-acceptance, and promoting awareness of digital distortion are crucial strategies for preserving positive mental health.

Fostering digital empathy emerges as an essential antidote. Recognizing that digital representations are selective and that everyone faces invisible challenges promotes a more understanding digital environment. By doing so, a more compassionate online community can be built, celebrating diversity and supporting personal growth.

In conclusion, understanding the constant quest for validation through online social comparison underscores the importance of fostering healthy digital practices and promoting a digital culture that nurtures authenticity and mutual support.

The constant comparison of oneself to others' seemingly perfect lives can lead to significant repercussions on self-esteem and self-image. In the vast digital landscape, where lives are presented with filters and curated narratives, understanding the psychological challenges that unfold in digital interactions becomes crucial.

The constant need for validation through likes, comments, and followers can become an addictive cycle. The absence of validation or the perception of a lack of positive interaction can profoundly impact self-esteem and contribute to depression, as personal worth is closely tied to digital response. Developing strategies for digital mental health becomes imperative to address anxiety and depression stemming from online social comparison.

Setting time limits on social media, practicing self-acceptance, and seeking support outside the digital environment are crucial steps to preserve mental health. Fostering open conversations about digital mental health breaks the stigma associated with online psychological challenges.

Creating spaces where users can share experiences, discuss digital pressure, and seek mutual support contributes to building a more compassionate online community. In summary, addressing anxiety and depression derived from online social comparison involves recognizing the challenges and actively working towards a digital culture that promotes authenticity, mutual support, and positive mental health.

Social validation in the form of likes, comments, and followers has become a digital currency fueling constant comparison. Online attention and approval are often perceived as direct indicators of personal worth, intensifying the need for constant validation in the digital world.

In the current digital landscape, likes, comments, and followers have transformed into a symbolic currency powering the constant comparison machine. Exploring the significance of this digital validation reveals how the quest for online attention and approval can shape the perception of personal worth, generating complex psychological dynamics in the vast digital landscape.

Social validation, represented by likes, comments, and followers, has emerged as a digital currency driving online interactions. The accumulation of this validation has been closely linked to the perception of personal worth, creating a cycle where digital attention becomes a direct marker of acceptance and recognition.

Online attention and approval are often perceived as direct indicators of personal worth. The quantity of likes and followers can be interpreted as a tangible measure of popularity and success, influencing self-esteem and generating the perception that one's value is intrinsically linked to digital response.

The importance of likes and digital validation fuels the incessant quest for online approval. The constant need to receive positive responses can drive specific behaviors, from carefully selecting content to actively engaging in trends, in an effort to maintain or increase perceived validation.

Digital validation can have a significant impact on self-image. The quantity of likes and comments can influence one's self-perception, generating a cycle where the lack of validation is interpreted as a sign of inadequacy. This can contribute to a negative self-image and the constant quest for validation to compensate for it.

The importance of likes also contributes to the dilemma of digital comparison. Constantly comparing with the validations received by others can generate feelings of competition and performance anxiety, as users seek to match or surpass the digital attention and approval of their peers.

Developing strategies for a healthy relationship with digital validation becomes essential. This includes practicing self-acceptance, understanding that online validation does not define personal worth, and setting limits on the importance given to digital response.

Promoting the recognition of personal achievements, beyond digital validation, is key. Valuing individual successes and authenticity over digital response contributes to building a strong self-esteem based on personal achievements and not simply on online approval.

In summary, understanding the importance of likes and digital validation highlights the need to cultivate healthy digital practices and promote an online culture that values authenticity and personal recognition beyond digital metrics.

The careful construction of a managed digital image can create additional stress. The pressure to maintain an impeccable online presence can be overwhelming, often leading to authenticity being sacrificed in favor of an idealized representation.

In the digital theater of social media, the careful management of the image can become a complicated dance between authenticity and the pressure to maintain a flawless online presence. Exploring the stress associated with the construction and maintenance of a digital image reveals the psychological tensions that can arise in the pursuit of digital perfection.

The construction of a flawless digital image entails constant pressure to maintain a perfect online presence. The competition for attention and digital validation can drive the need to project an image that resonates with digitally elevated standards, generating significant emotional burden.

Image management often comes with sacrificed authenticity. The pressure to meet expectations and receive validation can lead to the careful selection of moments to share, strategic content editing, and ultimately, the presentation of an idealized version of reality, leaving little room for true authenticity.

Image management becomes a complex dance between reality and representation. While attempting to balance the presentation of positive moments with authenticity, the line between truth and intentional projection often blurs, creating a constant tension between the need to please and the search for personal truth.

Digital expectations, driven by the social media culture, contribute to the stress associated with image management. The measure of success in terms of likes, comments, and followers can intensify the pressure to maintain an image that meets these expectations, generating anxiety and performance concerns in the digital realm.

The stress of image management can have a direct impact on mental health. The constant concern for maintaining a flawless image can generate anxiety, emotional exhaustion, and contribute to the feeling of not measuring up. The psychological burden associated with image management deserves careful attention.

Practicing digital authenticity emerges as a crucial strategy to mitigate the stress associated with image management. This involves allowing vulnerability, sharing real experiences, and embracing imperfection as an integral part of the digital narrative. Authenticity not only alleviates stress but also builds genuine connections.

Fostering a comprehensive digital culture is essential to alleviate the pressure associated with image management. Promoting digital empathy and recognizing that everyone faces invisible challenges contribute to building a more understanding online environment, where authenticity is valued beyond digital perfection.

In conclusion, recognizing and addressing the stress of image management involves a cultural shift towards more compassionate digital practices and the appreciation of authenticity as an invaluable digital asset.

Awareness of the negative impacts of online social comparison is the first step toward mitigation. Strategies such as mindfulness practice, setting time limits on social media, and focusing on personal development can help counteract the harmful effects.

Online social comparison can be an unrelenting force that veers the compass of self-esteem and emotional well-being. However, effective strategies exist to mitigate its negative impacts and foster a healthier relationship with the digital environment. These strategies are grounded in awareness, authenticity, and self-care. The first step toward mitigating social comparison is active awareness. Recognizing moments when comparison is experienced and understanding its negative effects is essential. Self-awareness provides the foundation for conscious decision-making and changing harmful thought patterns.

Mindfulness practice emerges as a powerful strategy. Cultivating mindfulness involves focusing on the present moment, reducing anxiety about future or past events that may fuel comparison. Mindfulness offers a mental refuge against the turbulent currents of constant comparison.

Setting time limits on social media is a proactive measure. Limiting the time spent on digital platforms not only reduces exposure to constant comparison but also allows for a more balanced connection between online and offline life.

The focus on personal development becomes a guiding beacon beyond comparison. Defining personal goals, cultivating skills, and working towards individual growth provide a solid foundation. Concentrating on one's own journey rather than comparing oneself to others promotes authenticity and personal satisfaction.

Fostering meaningful relationships outside the digital realm counters the feelings of isolation and competition. Establishing authentic and supportive connections outside of social media provides an emotional anchor, reminding that true validation and connection go beyond online likes and comments.

Valuing personal achievements beyond digital validation is essential. Recognizing and celebrating individual successes, no matter how small, strengthens self-esteem and detaches personal worth from digital metrics, thus mitigating the constant need for comparison.

Contributing to the creation of a positive digital culture is a shared responsibility. Fostering digital empathy, sharing real experiences, and celebrating the diversity of individual paths contribute to a more compassionate online environment and reduce the likelihood of destructive comparison.

In summary, these strategies to mitigate social comparison focus on self-awareness, authentic connections, and self-care, providing effective tools to address the psychological challenges of the digital environment.

Ultimately, online social comparison is a complex phenomenon that affects mental health and emotional well-being. Exploring these trends and their psychological consequences provides a foundation for addressing associated challenges and fostering a healthier relationship with digital platforms.

7.Cyberbullying and Mental Health

Cyberbullying, a dark shadow in the digital landscape, not only impacts online security but also has profound implications for mental health. Exploring the psychological impact of cyberbullying is essential to understanding the complexities of this phenomenon and developing effective strategies to combat it.

Cyberbullying can trigger significant anxiety and depression. Victims, facing threats and humiliations online, experience an overwhelming emotional burden, affecting their mental well-being.

Cyberbullying often leads to social isolation. Victims may withdraw from both online and offline interactions, fearing the continuation of bullying, contributing to loneliness and disconnection.

Constant aggression and criticism can erode the self-esteem of victims. The distorted perception of themselves, fueled by cyberbullying, may persist beyond digital screens.

Cyberbullying can also affect academic and professional performance. Constant distraction, emotional stress, and a loss of concentration can translate into difficulties at school or work.

Awareness and education are crucial. Creating widespread understanding of the effects of cyberbullying, as well as tools to prevent and address it, is essential for building a safer digital culture.

Building a safe digital culture requires a widespread understanding of the effects of cyberbullying and empowering the community to prevent and address this phenomenon. Establishing this awareness is a fundamental step toward creating a safer and more compassionate online environment.

Cyberbullying has a profound emotional impact on victims, causing anxiety, depression, and isolation. Understanding the severity of these consequences is essential to motivate preventive actions.

Cyberbullying, as a shadow in the digital world, not only leaves visible traces on the screen but also leaves a deep emotional impact on victims. Understanding the gravity of these consequences is essential to motivate preventive actions and cultivate a safer and more empathetic digital culture.

The resulting anxiety manifests as persistent digital distress. Victims may experience constant fear of new attacks, generating emotional hyperactivity in the face of the imminent threat of online harassment.

Constant hypervigilance characterizes anxiety in this context. Victims may feel the need to continually monitor their online profiles, anticipating and fearing new episodes of harassment.

The associated depression manifests in feelings of discouragement and hopelessness. Victims, affected by constant negativity online, may experience a loss of interest in previously pleasurable activities.

Self-esteem is significantly affected. It erodes the positive perception of oneself, contributing to a distorted and negative self-image that may persist beyond the digital realm.

Cyberbullying often leads to social isolation. Victims, fearful of online and offline interactions, may withdraw from their social environment, experiencing loneliness and disconnection.

Personal relationships can be affected, as victims may struggle to trust others due to experiences of online betrayal. It tends to distort social connections and the perception of trust.

Understanding the depth of the emotional impact creates intrinsic motivation to foster a culture of digital empathy. Empathy toward victims becomes a driving force for collective action. The gravity of the emotional impact highlights the importance of encouraging reporting. Creating awareness about the need to report is essential for victims to feel empowered and supported.

Recognizing the depth of emotional scars drives the need to develop support resources. Offering counseling and specialized psychological services becomes crucial to help victims overcome emotional trauma.

Understanding the profound emotional impact not only sheds light on the complexity of the phenomenon but also motivates preventive efforts and support actions. Digital empathy and collective awareness become essential tools.

Recognizing the long-term consequences of cyberbullying on mental health is essential to understand the severity of this phenomenon. From low self-esteem to increased anxiety, these repercussions not only affect the immediate emotional well-being of victims but also shape their long-term mental health, promoting empathy and motivating action.

Cyberbullying can erode the personal image of victims, contributing to low self-esteem. Constant online attacks and humiliations can make individuals doubt their worth and view themselves negatively.

Self-confidence is affected. Victims may experience a significant decrease in their ability to trust their own decisions and abilities, affecting not only their online presence but also their life beyond the screen.

This generates constant concern in victims. Anticipation of new attacks and uncertainty about who the perpetrators might be contribute to elevated levels of anxiety that can persist even outside the digital environment.

The resulting anxiety can affect daily life. Concentration on daily tasks may decrease, and victims may experience additional stress when facing situations that could trigger memories of the harassment.

Cyberbullying can contribute to the development of depressive disorders. Constant exposure to online hostility and the loss of a sense of security can immerse victims in a persistent mood of sadness and hopelessness.

Induced chronic anxiety can evolve into more severe anxiety disorders. Constant fear of new episodes of harassment can generate anxious thought patterns and avoidant behaviors.

Understanding the impact on mental health drives the need to foster digital empathy. Recognizing that victims may be dealing with significant emotional challenges motivates the online community to be more supportive and understanding.

Understanding the long-term consequences emphasizes the importance of preventive and support actions. From education to the creation of specific mental health resources, a comprehensive response is required.

In conclusion, recognizing that it can have long-term consequences on mental health is crucial to effectively address this issue. Empathy and collective action are essential to support victims and build a digital culture that prioritizes the emotional well-being of all users.

Cyberbullying also presents academic and professional challenges. Distraction and associated stress can negatively impact performance in school or work. Creating awareness of these impacts is essential for an informed response.

It is not limited to digital boundaries; it transcends the screen and infiltrates the academic and professional sphere. Academic and professional challenges stemming from cyberbullying, such as distraction and stress, can significantly impact academic and work performance. Creating awareness of these impacts is essential for an informed and compassionate response.

Victims may experience difficulties concentrating on their studies. Constant worry and distraction related to bullying can negatively affect the person's ability to focus on academic tasks.

Sustained lack of concentration can lead to lower academic performance. Victims may encounter additional challenges in completing assignments, participating in classes, and taking assessments, thus affecting their academic success.

Cyberbullying adds an additional layer of pressure on victims. The need to deal with online hostility while meeting academic expectations can generate significant stress, creating an additional burden for performance.

The associated stress can have a direct impact on emotional well-being. Constant anxiety and academic pressure can contribute to emotional issues that go beyond the school environment.

For those experiencing it in the workplace, the impact on productivity is significant. Dealing with online harassment can affect concentration and efficiency at work, harming the quality of work performance.

Professional relationships can also be affected. Lack of concentration and stress can influence the ability to interact effectively with colleagues, affecting workplace dynamics and collaboration.

Creating awareness involves educating about the academic and professional impacts of cyberbullying. Educational institutions and workplaces must implement awareness programs highlighting the severity of these challenges.

Providing academic and workplace support is crucial. Victims need resources to help them manage academic and professional challenges, whether through tutoring services, counseling, or flexibility at work.

In summary, recognizing the academic and professional challenges of the crime is essential to comprehensively address this issue. Education, support, and understanding in academic and workplace environments are key elements for building an informed and empathetic response to this phenomenon.

Education and awareness are powerful tools. Implementing educational programs that teach about its effects and how to prevent it contributes to a more informed culture.

Promoting the use of effective reporting platforms is essential. Victims should feel safe reporting, and platforms should respond quickly and efficiently to address the issue.

Teaching cybersecurity and privacy practices is crucial. Understanding how to protect personal information and use online security measures contributes to preventing cyberbullying situations.

Developing socioemotional skills is a valuable preventive tool. Teaching empathy, respect, and conflict resolution skills helps build healthier digital communities.

Facilitating access to psychological intervention and emotional support is essential. Victims need resources to help them overcome emotional impact, thereby strengthening their resilience.

Establishing clear legislation and policies is an important legal tool. Implementing significant consequences for harassers sends a clear message that such actions will not be tolerated.

Fostering a culture of digital empathy is essential. Empathy toward victims and the promotion of an online environment where respect and compassion prevail contribute to building safer communities.

Fostering a culture of digital empathy is essential to counteract it and cultivate safer and more compassionate online environments. Empathy toward victims, combined with the promotion of respect and compassion, forms the foundation for building healthier and more supportive digital communities.

Awareness begins with education. Online communities must understand the profound consequences, not only on an emotional level but also in terms of mental health, academic performance, and overall well-being.

Humanizing victims, emphasizing that behind every screen is a person with unique feelings and experiences, promotes a deeper connection and fosters empathy.

Establishing clear online behavior standards is essential. Defining what constitutes respectful behavior and what does not helps create a common standard that everyone should follow.

The digital community must actively condemn the crime. It is not just about not participating but intervening when necessary, reporting the behavior, and providing support to the victims.

Integrating socioemotional skills education in educational and digital environments helps develop empathy from an early age. Teaching users to understand and respect the emotions of others is fundamental.

Promoting positive and constructive online communication is essential. Digital communities should be places where opinions can be expressed without fear of harassment, promoting a respectful exchange of ideas.

Providing support resources for victims is crucial. From helplines to counseling services, ensuring accessible options for those affected is essential.

Implementing prevention programs that address the root of cyberbullying helps create more resilient communities. These programs can include workshops, talks, and awareness campaigns.

Digital empathy is based on collective responsibility. Involving the community in creating safe and respectful online environments strengthens the online social fabric and contributes to the prevention of this crime.

Leaders of digital communities, influencers, and authority figures must exercise exemplary leadership. Their actions and statements can have a significant impact on the overall digital culture.

In summary, fostering a culture of digital empathy is a collective task that requires continuous efforts and commitment. Empathy towards victims, the promotion of respect and compassion, along with education and support, are fundamental pillars for building safer and more compassionate digital communities.

Promoting inclusion and diversity online is key. A digital community that celebrates diversity and values each individual reduces the likelihood of discriminatory and hostile behaviors.

Promoting inclusion and diversity online is essential to build digital communities that celebrate the uniqueness of each individual. A digital culture that values diversity and fosters inclusion not only enriches the online experience but also reduces the likelihood of discriminatory and hostile behaviors.

Inclusion starts with representation. Digital communities should strive to represent diversity in all its forms, including, but not limited to, race, gender, sexual orientation, age, abilities, and cultural backgrounds.

Amplifying the voices of those who have traditionally been marginalized is crucial. Providing platforms for less heard voices contributes to more equitable representation.

Implementing awareness programs on diversity and inclusion in digital environments is essential. Education on cultural sensitivity and understanding others' experiences fosters a more respectful environment.

Demystifying stereotypes and prejudices through educational campaigns. Accurate and accessible information can challenge misconceptions and foster a genuine appreciation of diversity.

Establishing clear behavior standards that promote mutual respect. The digital community should be a space where everyone feels safe and valued, regardless of their differences.

Implementing concrete actions against discrimination. From zero-tolerance policies to reporting mechanisms, ensuring there are consequences for discriminatory behaviors is essential.

Facilitating moderated discussion forums that encourage constructive dialogue. These spaces should be safe environments where people can express their opinions respectfully and learn from others' experiences.

Promoting empathy by encouraging users to put themselves in others' shoes. Understanding different perspectives contributes to a more compassionate and collaborative digital culture.

Supporting financially and providing resources to initiatives that promote inclusion and diversity. This can include content projects, events, and educational programs that celebrate the richness of diverse experiences.

Collaborating with external communities and organizations dedicated to promoting diversity. Collaboration strengthens efforts and ensures a broader perspective in promoting inclusion.

Promoting inclusion and diversity online is imperative to build a digital culture where each individual feels respected, valued, and an integral part of the community. Celebrating the uniqueness of each voice enriches the online experience and contributes to the creation of fairer and more welcoming digital environments.

Building a safe digital culture involves education, prevention, emotional support, and the promotion of positive online values. Widespread awareness and community empowerment are the foundations for effectively addressing cyberbullying.

Promoting the importance of reporting and blocking is essential to empower users and effectively combat online harassment. Digital platforms must provide effective mechanisms that allow victims to report harassment and block perpetrators. Here are key elements to promote these crucial tools:

Launch educational campaigns to increase awareness of the importance of reporting and blocking. Users should understand that these actions not only protect their well-being but also contribute to the creation of safer online environments.

Share success stories where reporting and blocking have led to effective interventions. Highlighting how these tools can make a difference reinforces confidence in their effectiveness.

Design intuitive interfaces that facilitate the reporting process. Users should be able to easily identify and report harassing behaviors without complicated obstacles.

Provide detailed reporting options that allow victims to specify the nature of the harassment. This facilitates a more precise response from the platforms.

Ensure the confidentiality of reporters. Many victims may fear retaliation, so it is essential that the reporting process does not compromise their privacy.

Be transparent in the investigation process without revealing confidential information. Users should feel secure knowing that their reports are taken seriously.

Ensure that blocking is a quick and effective process. This allows victims to immediately end unwanted interaction and protect themselves from future harassment.

Provide clear notifications when someone has been blocked. This transparency helps victims take proactive measures to protect themselves.

Facilitate access to support resources after reporting. This can include links to helplines, counseling services, or online support communities.

Implement follow-up mechanisms to assess the ongoing well-being of victims after reporting. This demonstrates a long-term commitment to the safety and well-being of the community.

Promoting the importance of reporting and blocking not only empowers victims but also sends a clear message that harassment will not be tolerated in the online community. These tools are essential for building safe digital environments and protecting the online experience of all users.

Both victims and bystanders can benefit from resources addressing the emotional trauma caused by cyberbullying. Therapy and specialized counseling are valuable resources.

Offering psychological support is crucial to addressing the emotional trauma caused, benefiting both victims and bystanders.

Providing access to specialized counseling services for emotional trauma is crucial. Professionals trained in digital psychology can offer specific guidance to address the complexities of online harassment.

Offering options for both individual and group therapy. Individual therapy can be tailored to the specific needs of the victim, while group therapy can provide a sense of community and shared understanding.

Establishing online platforms dedicated to psychological support. These spaces can serve as safe communities where victims and bystanders share experiences, advice, and recovery resources.

Facilitating forums moderated by mental health professionals. The presence of experts ensures that the information provided is accurate, and interactions are positive and constructive.

Providing education on stress management techniques. Victims can learn practical strategies to cope with the emotional impact of cyberbullying and reduce anxiety.

Incorporating mindfulness and meditation practices into support resources. These techniques have proven effective in stress relief and promoting mental health.

Collaborating with mental health professionals specializing in digital psychology. These experts can bring specific insights into the psychological impact of cyberbullying and design tailored therapeutic approaches.

Offering training programs for mental health professionals addressing the uniqueness of related experiences. This ensures therapists are well-equipped to provide effective support.

Launching awareness campaigns to promote psychological support resources. Ensuring that victims and bystanders are aware of and have access to these services is essential for their effectiveness.

Working to reduce the stigma associated with seeking psychological support. Fostering an environment where asking for help is seen as a brave and healthy act.

Providing effective psychological support is an integral part of addressing cyberbullying and its emotional consequences. By creating a comprehensive and accessible support system, significant contributions are made to the healing and empowerment process of victims and bystanders.

Strengthening policies and legislation is essential to combat this form of digital violence and ensure a safer online environment. Establishing legal consequences for harassers not only seeks accountability but also acts as an effective deterrent.

Clearly defining cyberbullying, including various forms such as verbal harassment, social media harassment, online defamation, and non-consensual sexting. This provides a solid foundation for addressing a variety of harmful behaviors.

Considering age in the definition to address cyberbullying specifically targeting children and adolescents. Policies must adapt to effectively address age-related dynamics.

Establishing clear legal consequences for harassers, including sanctions and penalties proportional to the severity of cyberbullying. This sends a strong message that these actions will not be tolerated.

Incorporating measures to protect victims during and after legal processes. This may include digital restraining orders and additional security measures to safeguard the well-being of those affected.

Implementing mandatory reporting policies for online platforms. Requiring platforms to report cases to relevant authorities ensures a swift and coordinated response.

Encouraging effective collaboration between online platforms and legal authorities. This can include secure transfer of digital evidence and cooperation in investigations.

Implementing educational programs on cybersecurity and digital ethics. Prevention is crucial, and education can help reduce the incidence of the crime by increasing awareness of its impacts and legal consequences.

Improving public awareness of existing legislation against cyberbullying. This can include informational campaigns to educate society about their rights and available resources.

Conducting regular policy reviews to adapt to technological advancements and new forms in this criminal field. Legislation must stay up-to-date to effectively address changing dynamics online.

Collaborating with technology experts to understand and address the complexities of crime in constantly evolving digital environments.

Strengthening policies and legislation is an essential component of creating a safer online environment and protecting people from digital violence. These measures must be comprehensive, adaptive, and supported by joint efforts between governments, online platforms, legal professionals, and society at large.

Fostering digital empathy is key. Educating about the real consequences and promoting understanding towards victims contributes to building more supportive online communities.

Fostering digital empathy is essential to create more supportive and resilient online communities. By educating about the real consequences of these actions and promoting understanding towards victims, a more empathetic and collaborative online environment can be cultivated.

Including digital empathy in school curricula. Educational programs can address topics such as the emotional impact of cyberbullying, online responsibilities, and how to support those who have been affected.

Organizing workshops to raise awareness about digital empathy. These events can involve students, parents, and educators to create a shared understanding of the importance of online empathy.

Sharing impactful stories that highlight the real consequences of cyberbullying. These campaigns can be driven by government organizations, NGOs, and technology companies to reach diverse audiences.

Using social media platforms to spread messages of digital empathy. Viral campaigns can have a significant impact by engaging a wide audience in the conversation.

Developing interactive simulations that allow users to virtually experience the consequences. This can enhance understanding by providing a more personal perspective.

Creating multimedia resources, such as videos and podcasts, addressing digital empathy. Combining formats can reach different types of audiences and maintain interest.

Involving public figures and community leaders in digital empathy campaigns. Their participation can increase visibility and credibility of the messages.

Establishing partnerships with educational institutions, non-profit organizations, and companies to support digital empathy initiatives. Collaboration can amplify impact and reach diverse audiences.

Facilitating open discussion forums on online platforms. These spaces allow people to share experiences, ask questions, and receive support, contributing to the creation of more understanding online communities.

Implementing active moderation to ensure that forums are safe and respectful environments. This fosters healthy communication and prevents the spread of harmful behaviors.

Fostering digital empathy not only contributes to preventing cyberbullying but also strengthens cohesion and compassion within online communities. These strategies aim to create a cultural shift towards a more conscious and empathetic digital culture.

Working towards creating safe digital environments is a collective effort. Platforms, educational institutions, and society at large must collaborate to implement measures that prevent and address cyberbullying.

Creating safe digital environments is an imperative that requires collaboration from online platforms, educational institutions, and society at large. This joint effort is essential to effectively prevent and address the issue. Here are key strategies for creating safe digital environments:

Develop and clearly communicate online conduct guidelines. This includes specific guidelines on respect, empathy, and the consequences of such actions on digital platforms.

Implement zero-tolerance policies for cyberbullying on all platforms. This ensures that harassers face immediate consequences and provides a safer environment for users.

Establish strong partnerships with educational institutions. Collaboration between digital platforms and schools can ensure a coordinated response and ongoing education on cybersecurity.

Involve the community in creating safe digital environments. Active participation from parents, educators, and community leaders is crucial to addressing the crime from multiple perspectives.

Provide support resources for victims and witnesses. These resources may include helplines, psychological counseling, and intervention programs to mitigate emotional impact.

Encourage the creation of online support communities. Support groups can offer a safe space for affected individuals to share experiences and receive guidance.

Creating safe digital environments is an ongoing process that requires constant commitment from all involved parties. By joining forces, digital platforms, educational institutions, and society at large can work towards a safer and more positive online environment for everyone.

In conclusion, addressing cyberbullying and its impact on mental health requires a comprehensive approach that combines education, emotional support, and legislative action. Creating a safe and compassionate online environment is essential to preserve the mental health of those navigating digital complexities.

8. Virtual Reality and Empathy

The convergence of virtual reality (VR) and empathy has generated a fascinating field of possibilities. By providing immersive experiences, virtual reality has the potential to transform the way we perceive and connect with others' experiences.

Virtual reality allows users to experience the world from different perspectives. By immersing themselves in specific environments and situations, users can better understand the realities faced by those with unique challenges, fostering empathy by changing their own points of view.

The facilitated shift in perspectives represents a revolutionary ability to transform our understanding of the world and others' experiences. By immersing users in specific environments and situations, virtual reality goes beyond mere observation, allowing them to immersively experience the realities of those facing unique challenges. This process creates a deeper emotional connection by bringing about a significant shift in individual perspective.

Empathy is often hindered by the inability to fully comprehend others' experiences. Virtual reality eliminates these barriers by providing a platform where users can literally walk in someone else's shoes. This deep immersion overcomes the limitations of traditional empathy, generating a more complete and nuanced understanding.

Experiencing the world from different perspectives through virtual reality not only transforms individuals but also has the power to generate social awareness on a large scale. Immersive simulations can address critical social issues, such as discrimination or lack of access to basic resources, vividly illustrating the daily struggles faced by many individuals worldwide.

By changing perspectives, virtual reality contributes to increased tolerance and understanding among different groups of people. It allows users to experience the diversity of human life in a way that goes beyond theory, fostering a deeper appreciation for cultural, social, and economic differences.

Interpersonal empathy is strengthened when users can live through others' experiences via virtual reality. This tool not only educates about diverse realities but also evokes genuine emotions and sensations, establishing a richer and more authentic connection between people.

In the educational realm, it has become an invaluable tool for expanding students' understanding. From virtual trips to historical eras to simulations of cultural contexts, this technology transforms learning by enabling students not only to learn about diverse realities but also to actively experience them.

Ultimately, the facilitated shift in perspectives by virtual reality represents a revolutionary advancement in how we relate to the world around us. By breaking down empathy barriers and offering immersive experiences, this technology becomes a catalyst for building bridges of understanding and solidarity in global society.

Through immersive simulations, VR enables users to live through others' experiences. This can range from everyday situations to extraordinary events, creating a deeper emotional connection by directly experiencing the emotions and challenges of others.

Living through others' experiences via immersive simulations (VR) is a powerful way to emotionally connect users with the experiences of other people. VR's unique ability to recreate environments and events offers a virtual window into different worlds, allowing users to fully immerse themselves in situations that can vary from the ordinary to the extraordinary.

Virtual reality goes beyond mere observation; it allows users to be active participants in narratives that are not their own. By living through others' experiences, users not only intellectually understand the emotions and challenges of others but also experience those events directly. This deep immersion intensifies emotional connection and promotes more authentic empathy.

VR provides a versatile platform to explore human diversity. From the everyday life of different lifestyles to the extraordinary complexity of unique events, users can experience the breadth and depth of human experiences. This contributes to the appreciation of individual and cultural differences.

Living through others' experiences, users become virtual witnesses to the realities that others face. This impact on social awareness can be powerful, fostering a greater understanding of global issues and promoting a sense of collective responsibility to address shared challenges.

VR enables the creation of immersive narratives that go beyond simple visual representation. By integrating sensory elements such as sound and touch, experiences become even more convincing. This authenticity in representation contributes to a deeper emotional connection with the stories being told.

Living through others' experiences via VR can address intersectionality by allowing users to understand the complex intersections of identities and experiences. This contributes to a more complete empathy by recognizing the interconnections of factors such as gender, race, class, and sexual orientation in shaping individual experiences.

Experiencing others' situations in controlled environments can have therapeutic applications. From phobia treatments to building empathy in therapies, VR offers tools to explore and address a variety of emotional and psychological challenges.

Living through others' experiences via immersive simulations not only pushes the boundaries of technology but also transforms how we relate to human narratives. By directly experiencing the emotions and challenges of others, users can cultivate a deeper connection with the diversity of the human experience.

VR has been integrated into educational environments to create simulations that foster empathy. This is especially valuable in fields such as medicine and psychology, where students can practice and understand patients' experiences in a realistic yet safe way.

The integration of virtual reality (VR) into educational environments has marked a significant milestone by providing students with immersive experiences that foster empathy. In fields such as medicine and psychology, where a profound understanding of patients' experiences is crucial, it has become a valuable tool.

It allows the creation of realistic simulations that replicate real-world situations. In educational environments, this means that students can practice and understand patients' experiences in a controlled and safe environment. For example, future medical professionals can perform virtual procedures before facing real situations.

Empathy is essential in fields such as medicine, where professionals must understand and respond to patients' emotional needs. VR provides opportunities for students to immerse themselves in patients' perspectives, thereby developing crucial empathic skills for quality care.

In medical and mental health settings, it facilitates training in clinical skills. Students can practice handling delicate situations, such as difficult diagnoses or sensitive conversations with patients, improving their ability to deal with complex emotional scenarios.

Virtual environments allow students to understand patients' experiences from various perspectives. For example, in psychology, they can immerse themselves in environments simulating mental disorders, contributing to a deeper understanding of the experiences of those facing mental health challenges.

VR benefits not only students but is also used in ongoing medical education. Practicing healthcare professionals can stay updated and hone their skills through virtual simulations, enhancing the quality of care they provide throughout their careers.

In educational settings, it promotes an interdisciplinary approach by allowing students from different fields, such as medicine, psychology, and nursing, to collaborate in virtual simulations. This reflects the reality of healthcare, where collaboration among professionals is essential.

It facilitates effective assessment and feedback. Educators can monitor and analyze students' performance in virtual environments, providing specific and personalized feedback to enhance their competence and empathic understanding.

The integration of virtual reality into educational environments not only enriches students' training but also improves the quality of healthcare and mental health care by cultivating essential empathic skills in future professionals.

Immersive narratives in VR can transport users to specific cultural, historical, or social contexts. These stories enable the exploration of different realities, promoting understanding and empathy towards diverse life experiences.

Immersive stories have opened a new dimension in storytelling, allowing users to fully immerse themselves in specific cultural, historical, or social contexts. These experiences not only entertain but also promote understanding and empathy towards diverse life experiences. Key aspects of immersive stories in VR are highlighted here:

Immersive narratives have the power to transport users to places and historical moments in a completely engaging way. Whether exploring ancient civilizations, reliving historical events, or delving into distant cultures, it creates a unique bridge between the user and the content.

It allows users to interact and participate in narratives, fostering a deeper understanding of diverse cultures. The ability to actively explore cultural environments contributes to a more authentic and respectful appreciation.

By immersing themselves in the experiences of characters and communities through immersive stories, users build bridges of empathy. This emotional connection is strengthened by closely experiencing the challenges, achievements, and narratives of those who are different, fostering greater understanding and tolerance.

Immersive stories amplify narrative diversity by offering experiences beyond traditional storytelling. Stories told from less represented perspectives can challenge stereotypes and provide a platform for voices often overlooked in mainstream media.

Immersive stories not only inform but also educate through experiential learning. Users can learn about historical events, cultural traditions, or social challenges by virtually living those experiences, enhancing knowledge retention and deep understanding.

By directly experiencing life in specific cultural contexts, users develop a sharper cultural sensitivity. This is especially valuable in an increasingly interconnected world where understanding and respect for diversity are crucial.

Immersive stories can become a platform for global empathy. By connecting with experiences from people worldwide, users can develop a deeper appreciation for the richness and complexity of the human condition, transcending geographical and cultural boundaries.

In summary, immersive stories in virtual reality open new frontiers in storytelling by offering experiences that not only entertain but also educate and foster empathy. By immersing users in diverse contexts, these narratives contribute to cultural enrichment and the building of a more understanding world.

In the field of mental health, it is used in exposure therapy. This involves creating safe environments for individuals to gradually face their fears and anxieties, offering a controlled way to address traumas and phobias.

Exposure therapy has proven to be an innovative and effective tool in the field of mental health. This technique involves creating safe and controlled virtual environments for individuals to gradually face their fears, anxieties, or traumas.

It provides safe and controlled virtual environments where individuals can confront situations that cause them anxiety or fear. These environments are customizable and adaptive, allowing therapists to adjust the intensity of exposure according to each patient's specific needs.

Exposure therapy facilitates the gradual addressing of phobias. Therapists can design scenarios that represent specific triggers for phobias, enabling patients to progressively and controlledly confront those fears. This gradual approach promotes habituation and reduces associated anxiety.

It has been successfully used in the treatment of various anxiety disorders, such as post-traumatic stress disorder (PTSD), social anxiety disorders, and specific phobias. Therapeutic exposure in VR provides individuals with an additional tool to cope with and overcome challenges associated with these disorders.

Each individual has unique experiences and triggers, and VR allows for the customization of treatment. Therapists can tailor virtual environments to address each patient's specific concerns, ensuring a more effective and individual-focused approach.

VR offers a level of immersion and realism crucial for the success of exposure therapy. By immersing individuals in convincing virtual environments, the treatment's effectiveness is increased by faithfully replicating real-life situations that induce anxiety.

VR provides access to environments that might not otherwise be available for traditional exposure therapy. For example, recreating specific childhood environments or simulating challenging situations can be essential for addressing past traumas or deeply rooted phobias.

Its use can reduce the stigma associated with certain mental disorders, as therapeutic exposure takes place in a virtual environment. This may contribute to greater acceptance by individuals seeking treatment, as the experience can feel more discreet and less intrusive.

Exposure therapy represents a valuable tool in the treatment arsenal for anxiety disorders and phobias. By providing customizable and safe virtual environments, it offers an effective and adaptable way to address mental health challenges.

VR has also been employed to foster emotional connections in therapies. By creating virtual environments where users can express their emotions more freely, the therapeutic process is facilitated, and therapist-patient relationships are strengthened.

Its use to foster emotional connections in therapies represents a significant advancement in the field of mental health. By creating virtual environments that encourage the free expression of emotions, the therapeutic process is enhanced, and relationships between therapist and patient are strengthened.

It offers the possibility to design virtual environments specifically aimed at facilitating emotional expression. These environments can range from serene landscapes to interactive scenarios that allow users to visually represent their emotions in ways that may be challenging in the traditional therapy setting.

The immersive nature of VR provides patients with a space where they can express their emotions more freely and openly. By removing some perceived barriers, such as shame or fear of judgment, patients can delve more deeply into the exploration and communication of their feelings.

For therapists, it can provide a unique window into patients' emotions. By observing how patients interact with the virtual environment, therapists can gain a deeper and more empathetic understanding of their emotional experiences, informing and enriching therapy sessions.

The customization of therapeutic scenarios allows tailoring the experience to the individual needs of each patient. Creating environments that resonate with each person's specific emotional experiences and challenges contributes to more effective and patient-centered therapy.

It can integrate various therapeutic techniques, such as exposure therapy or play therapy, into virtual environments. This provides therapists with a versatile tool to address a variety of emotional and psychological issues.

The facilitated emotional connection can strengthen the therapeutic relationship. By experiencing virtual environments and situations together, therapist and patient can build a deeper connection, enhancing trust and collaboration in the therapeutic process.

For individuals who have experienced trauma, it can offer a more gradual and controlled therapeutic approach. Therapists can use virtual environments to address trauma in a carefully guided manner, promoting healing and the management of associated emotions.

In conclusion, the application of virtual reality to foster emotional connections in therapies represents an exciting innovation in the field of mental health. By leveraging VR's immersion and customization, a new path is opened to enrich therapeutic experiences and promote greater emotional well-being.

VR has been used in awareness campaigns to highlight social issues. By transporting participants to challenging situations, these campaigns aim to generate empathy and raise awareness about issues such as poverty, homelessness, and discrimination.

Its use in awareness campaigns to address social issues is a impactful strategy that seeks to generate empathy and awareness about important issues. Key aspects of how VR has been employed in this context are explored here:

VR allows participants to be virtually transported to challenging situations that reflect the challenges associated with specific social issues. Whether experiencing life on the streets, battling poverty, or facing discriminatory situations, it provides a deep immersion that goes beyond mere visual understanding.

By enabling participants to virtually live the experiences associated with social issues, VR seeks to generate empathy in a more direct and visceral way. The immersive experience can evoke deeper emotional responses, leading to a more comprehensive understanding of the realities faced by affected individuals.

Awareness campaigns employ VR as a powerful tool to raise awareness about social issues. By virtually experiencing these problems, participants can gain a more authentic perspective, leading to increased awareness and understanding of the complexities involved.

VR allows participants not only to be observers but also active participants in the narrative. This active engagement contributes to a deeper connection with the stories and experiences presented, fostering more meaningful engagement with the highlighted cause.

The immersion provided by VR has the potential to influence the decisions and behaviors of participants. By directly experiencing the challenges associated with social issues, individuals may feel a stronger motivation to advocate for change and engage in concrete actions to address these problems.

Using technology to amplify the message of awareness campaigns, VR experiences become more accessible, reaching a broader audience. This includes individuals who might not have otherwise been aware or engaged with the highlighted social issues.

he successful implementation of campaigns often involves close collaboration with social organizations and advocates. This collaboration ensures that virtual representation is accurate, respectful, and aligns with the goals and values of the affected communities.

The application of virtual reality in awareness campaigns addressing social issues aims to leverage technology to generate a deeper understanding and emotional connection with the realities faced by many communities. By offering immersive experiences, these campaigns aspire to drive social change and advocate for more effective solutions.

Experiencing the realities of those facing challenges through VR can have a significant impact on social awareness. VR becomes a powerful tool for mobilizing support and action around various causes.

The ability to virtually experience the realities of those facing challenges through VR has a significant impact on social awareness.

VR provides an immersive experience that goes beyond simple visual information. By allowing participants to virtually experience the realities of those facing challenges, empathy and emotional connection are generated. This emotional connection is crucial for fostering a deeper awareness of the struggles and difficulties faced by affected individuals.

The immersive experience creates a lasting impact on participants' memory and perception. Virtual experiences have the potential to leave a deeper and more memorable impression than other forms of information or awareness. This lasting impact contributes to keeping social awareness alive over time.

By virtually experiencing the realities of those facing challenges, participants are more likely to feel an internal motivation to act. VR becomes a mobilizing tool that inspires people to actively engage, whether through donations, volunteering, or advocating for policies that address the presented issues.

It facilitates active participation by allowing participants not only to see but also to live the presented experiences. This active engagement contributes to the development of a sense of responsibility and commitment to social causes, encouraging people to take an active role in creating positive change.

With increasingly accessible technology, these experiences can reach broader audiences. This includes people who may not have had direct access to information or experiences related to the highlighted social issues. VR extends the reach of social awareness, building bridges between diverse communities and cultures.

It contributes to the building of supportive communities by bringing together people with a common interest in addressing specific social issues. Shared experiences can foster collaboration and solidarity, creating support networks that work together to achieve positive impact.

VR stands out for its ability to educate through experiences. By offering an authentic and experiential view of social issues, it effectively educates, overcoming barriers of indifference or lack of understanding.

In conclusion, virtual reality emerges as a powerful tool for social awareness by offering immersive experiences that evoke empathy, inspire action, and build communities committed to creating positive change in the world.

It offers interactive simulations that allow users to practice empathic skills in controlled environments. This is particularly beneficial in fields such as customer service, where empathy is crucial.

The ability to provide interactive simulations has led to significant benefits, especially in fields where empathy plays a crucial role, such as customer service.

Interactive simulations enable users to actively train and improve their empathic skills. By immersing themselves in simulated environments, participants can practice empathy in realistic and controlled situations, developing competencies that are transferable to real-world contexts.

It provides controlled and safe environments for the practice of empathic skills. This is especially beneficial in situations where hands-on learning might be challenging or risky. Participants can make mistakes and learn from them in a virtual environment before facing real-life situations.

Programs can customize scenarios based on the specific needs of the training. This allows addressing particular situations relevant to customer service, adapting to the challenges and requirements specific to each industry or company.

It facilitates the delivery of immediate feedback. Participants can receive detailed assessments of their empathic performance, allowing them to adjust their approach and continuously improve. This immediate feedback accelerates the learning process.

The immersive nature contributes to better information retention. Experiences lived in virtual environments tend to leave a more lasting impression, making it easier to transfer empathic skills acquired during training to real-world situations.

Interactive simulations not only address surface-level empathy but also contribute to the development of emotional intelligence. By facing emotionally charged situations in virtual environments, participants learn to better recognize, understand, and manage emotions, enhancing their overall emotional competence.

Programs can simulate challenging scenarios that are difficult to replicate in reality, allowing users to practice empathy in extreme or unusual situations. This prepares customer service professionals to deal with a variety of circumstances.

While the initial investment in technology may be significant, interactive simulations can be more efficient and cost-effective than traditional training methods in the long run. Costs associated with training in real-world scenarios are eliminated, and the effectiveness of learning is maximized.

In summary, interactive simulations offer a valuable tool for the development of empathic skills in customer service and other fields. The ability to practice in controlled environments, receive immediate feedback, and adapt to diverse scenarios makes it an effective option for enhancing empathy and emotional intelligence in a practical and efficient manner.

Training programs based on it can provide immediate feedback, allowing users to adjust and improve their empathic responses. This contributes to the continuous development of emotional skills.

The use of virtual reality in training programs offers a valuable tool for the continuous improvement of emotional skills, especially in the context of immediate feedback and adjustment.

Training programs allow for real-time feedback. Users receive instant assessments of their empathic responses, enabling them to immediately understand the effectiveness of their actions and communication in simulated situations.

Immediate feedback facilitates the identification of specific areas that require improvement. Participants can analyze their virtual interactions and understand where they can adjust their approach to strengthen their emotional skills, from empathy to conflict management.

Based on the feedback received, programs can customize learning for each user. This means adapting scenarios and specific challenges that address the identified areas of improvement, maximizing the effectiveness of training.

Immediate feedback allows users to practice iteratively. They can make adjustments and re-encounter similar situations to consolidate and improve their emotional skills over time. This iterative practice approach is essential for the sustainable development of skills.

Feedback acts as a stimulus for reflection. Participants can review their interactions and reflect on their performance, fostering a deeper understanding of how their actions affect the emotional dynamics of a situation.

Since feedback can be tailored to different learning styles, programs are versatile. Some users may benefit more from visual cues, while others may prefer verbal feedback. Customization ensures that feedback is effective for each individual.

Immediate feedback serves as a motivator for continuous improvement. Users can directly see how their adjustment efforts positively impact their emotional skills, fostering ongoing engagement with development and evolution.

VR systems can keep detailed records of each user's progress. This allows for continuous monitoring and the ability to assess progress over time. Progress records contribute to accountability and transparency in the learning process.

Training programs based on it not only offer immersive experiences for the development of emotional skills but also leverage immediate feedback to drive continuous improvement. The combination of iterative practice, customization, and stimulus for reflection contributes to the sustainable growth of participants' emotional competencies.

Virtual reality emerges as a transformative tool for building stronger human connections and fostering empathy.

9.Notifications and Anxiety

The impact of constant notifications on mobile devices on anxiety and stress is a relevant topic in the digital era. They can contribute to information overload. The avalanche of information, from emails to social media updates, can induce anxiety by making individuals feel overwhelmed and disconnected.

Information overload, exacerbated by constant notifications, represents a significant challenge in the digital era. They fuel an uninterrupted flow of information. From emails and text messages to app alerts, individuals are immersed in a constant stream of data that can be overwhelming.

The multiplicity of sources and messages complicates the task of prioritizing information. Individuals may feel the pressure to keep up with everything, leading to a sense of overwhelm when trying to process and assimilate an excessive amount of data.

Constant exposure to information can lead to information fatigue. This mental exhaustion manifests when people feel saturated, impacting their ability to process new information effectively.

The feeling of always being connected can generate anxiety, as individuals may struggle to detach from digital information even during moments of rest or recreation.

Information overload creates a constant pressure to respond immediately. This expectation of perpetual availability can generate anxiety, as individuals feel compelled to stay active and respond in real-time.

Information saturation can also contribute to social disconnection. When people are constantly immersed in their devices, they may lose significant connections with their environment and interpersonal relationships, creating a sense of isolation.

Information overload can negatively impact productivity. Difficulty in filtering relevant from irrelevant information can lead to a decrease in efficiency and concentration on specific tasks.

Handling information overload requires effective filtering skills. The ability to discern what information is relevant and valuable is essential to reduce anxiety associated with the overwhelming amount of data.

Developing strategies to manage information overload is crucial. Establishing time limits, using filtering tools, and practicing digital disconnection are effective approaches to mitigate anxiety derived from the overwhelming amount of information.

The information overload generated by constant notifications can have significant consequences on mental health and well-being. Addressing this challenge involves developing management skills and adopting balanced approaches to maintain a healthy relationship with digital information.

Frequent interruptions due to notifications can affect workflow and concentration. The need to constantly shift attention from one task to another can create stress, especially in work or academic environments.

It represents a significant challenge in work and academic environments, impacting concentration and productivity.

Frequent notifications can disrupt the natural flow of concentration. Shifting attention from one task to another due to digital alerts can hinder deep immersion in a specific activity.

The need to constantly switch between tasks can result in decreased efficiency. The mind requires time to adjust to the context of a new task, leading to a loss of time and energy.

Frequent interruptions can generate stress and anxiety in the workplace. Professionals may feel the pressure to meet deadlines and expectations, and constant notifications can contribute to the feeling of overwhelm.

The quality of work can be negatively affected. Lack of continuous focus due to digital interruptions can lead to errors and less attention to detail.

Receiving notifications can trigger a cycle of continuous interruption. Each alert can lead to the review of multiple applications and platforms, prolonging the disconnection from actual work.

After an interruption, restoring the original focus on the task may take time. This "restoration" process adds an additional layer of complexity to the workflow.

Recognizing the importance of interruption-free environments is crucial. Establishing specific times for checking notifications and, at the same time, reserving periods of uninterrupted work can improve productivity and reduce stress.

Developing effective time management strategies becomes essential. Establishing dedicated time blocks for specific tasks and limiting digital interruptions can contribute to a more effective workflow.

Organizational culture also plays a crucial role. Promoting work practices that value concentration and minimize distractions can significantly improve productivity and well-being in the workplace.

Addressing workflow disruption due to constant notifications involves adopting proactive approaches to manage attention and promoting environments that support concentration and productivity.

Notifications create expectations of immediate response. This can generate anxiety, as people feel the pressure to always be available and respond quickly to messages, emails, or updates, even outside of working hours.

The creation of expectations of immediate response is a significant consequence of constant notifications, and its impact spans various areas.

They generate an implicit pressure to always be available. People may feel obligated to respond even outside their working hours, contributing to a culture of constant availability.

The expectation of immediate response can blur the boundaries between personal and professional life. The ability to receive notifications on personal devices can make work demands seep into moments intended for rest and personal activities.

The pressure to respond instantly can impact mental health. It contributes to stress, anxiety, and a constant sense of alertness, as people feel the need to meet response expectations.

The constant anticipation of notifications can lead to a cycle of constant responsiveness. People may feel the need to regularly check and respond to notifications, even in situations where disconnecting and resting would be beneficial.

Expectations of immediate response can make it difficult for people to disconnect from work. The feeling of always being "on call" can interfere with the ability to enjoy leisure time and relax.

Constant attention to notifications can also affect personal relationships. Continuous distraction can interfere with the quality of time spent with friends and family, contributing to interpersonal disconnection.

To manage expectations of immediate response, it is essential to set clear boundaries. This involves effectively communicating availability schedules and fostering an organizational culture that values the balance between work and personal life.

Organizations can contribute to the mental health of their employees by promoting healthy practices regarding notifications. This may include policies that respect time outside of work and encourage digital disconnection during certain periods.

Education on digital time management can be beneficial. Providing people with tools and strategies to manage notifications and set boundaries can improve quality of life and reduce associated anxiety.

Overall, addressing expectations of immediate response requires reflection at both the personal and organizational levels to establish healthy boundaries and promote practices that support mental well-being and a balance between work and personal life.

Constant notifications are also linked to digital FOMO, where people fear missing out on events or relevant information. The constant need to stay informed can generate anxiety about the possibility of missing something important.

Digital FOMO (Fear of Missing Out) is a specific manifestation of the fear of missing events or relevant information in the digital environment, and constant notifications play a key role in intensifying this phenomenon.

Notifications act as constant stimuli that keep people connected to events and real-time updates. This constant exposure can fuel digital FOMO by creating a sense of urgency and exclusivity around the information being shared.

The constant arrival of notifications creates continuous expectations of novelty and relevant information. This can generate anxiety, as people fear missing something important if they don't respond or check notifications immediately.

Digital FOMO is inherently linked to social anxiety in a digital environment. Constant exposure to seemingly exciting events and activities can lead to constant comparison and worry about not participating in similar experiences.

Information in the digital environment is often ephemeral, with rapid updates and relevance expiration. Notifications reinforce the idea that it is crucial to be aware of novelties at the exact moment they occur, increasing anxiety about missing something important.

Notifications also contribute to the cycle of instant validation. The possibility of receiving likes, comments, or immediate interactions in response to a post can fuel the desire to constantly engage to avoid the feeling of falling behind.

The constant arrival of notifications can create an implicit pressure to always be connected. People may feel the need to check their devices frequently to avoid missing events or updates, even when they should be disconnected.

To mitigate the impact, it is essential to set limits on the use of notifications. Establishing specific time periods for disconnecting and actively managing expectations regarding information availability can help reduce anxiety associated with the fear of missing out.

Promoting a more balanced understanding of online information and its effects can help people cope more healthily with anxiety associated with digital FOMO.

Addressing digital FOMO involves recognizing the complex interaction between notifications, the need for instant validation, and anxiety about missing experiences, promoting strategies that encourage a healthier and more balanced relationship with digital technology.

Nighttime notifications can negatively affect sleep. Receiving messages or alerts during the night can disrupt rest, contributing to fatigue and long-term anxiety.

The impact of nighttime notifications on sleep is a crucial aspect to consider in the context of mental health and overall well-being.

Nighttime notifications have the potential to disrupt the natural sleep cycle. Receiving messages or alerts can wake people up at night, preventing them from reaching the deeper and more restorative stages of sleep.

It can generate anxiety, as people may feel the pressure to respond or address messages, even when they should be resting. This anxiety can contribute to a state of wakefulness and hinder the relaxation necessary for falling asleep.

Sleep disruption due to nighttime notifications can result in daytime fatigue. The lack of restful sleep is commonly associated with a decrease in cognitive and physical performance, negatively affecting quality of life.

Disordered sleep is linked to mental health problems, such as anxiety and depression. Nighttime notifications, by contributing to sleep disruption, can exacerbate these issues and impact long-term mental health.

To preserve sleep quality, it is essential to set limits on nighttime notifications. Configuring "do not disturb" modes during certain hours or placing devices out of reach during the night are effective strategies to minimize sleep disruption.

Educating about the importance of healthy sleep habits and the need to disconnect before bedtime is fundamental. Promoting relaxing routines before sleep and creating a conducive environment for rest help counteract the negative effects of nighttime notifications.

Awareness of digital health includes recognizing how notifications impact overall well-being. Promoting the importance of uninterrupted and restorative sleep contributes to a broader understanding of technology's impacts on health.

Managing nighttime notifications is essential to safeguard sleep and promote optimal well-being. Establishing clear boundaries and encouraging healthy sleep habits are crucial steps in mental and physical health care.

Notifications are often linked to social pressure and digital validation. Instant response to notifications can become a constant quest for validation, generating anxiety about how others perceive oneself.

The relationship between notifications, social pressure, and digital validation is a relevant phenomenon in the digital era. Key aspects of this dynamic are explored here:

Receiving notifications often creates expectations of an immediate response. The pressure to always be available to respond to messages or interact on social media can generate anxiety, as people feel obligated to meet these social expectations.

Notifications, especially on social media platforms, are closely linked to the search for validation. Positive responses in the form of likes, comments, or interactions are perceived as social validation, fueling the constant need for this type of digital recognition.

The connection between notifications and digital validation can significantly impact self-esteem. Lack of interaction or the absence of positive responses can create doubts about one's own worth and contribute to a negative self-image.

Social pressure and digital validation through notifications often lead to constant comparison. People may measure their worth based on the quantity and quality of online interactions, generating a cycle of comparison that negatively affects mental health.

Anxiety about how others perceive oneself intensifies with the response to notifications. The speed of response and the quantity of interactions received can become indicators of popularity and acceptance, contributing to social anxiety.

The relationship between notifications and digital validation can create a constant need for validation. People may become dependent on positive online feedback, continually seeking approval from others to reaffirm their worth.

Recognizing social pressure linked to notifications and digital validation is the first step toward healthy management. Practicing digital disconnection, setting boundaries, and fostering self-acceptance are key strategies to mitigate the negative effects of this dynamic.

The interaction between notifications, social pressure, and digital validation can have a profound impact on mental health. Fostering a balanced relationship with technology and promoting self-acceptance regardless of online interactions are essential elements for emotional well-being.

Constant stimulation through notifications can contribute to digital addiction. The compulsive need to check the phone or device for new notifications can generate anxiety if digital connection becomes a difficult habit to break.

Digital addiction, characterized by the compulsive dependence on devices and online services, finds a crucial component in notifications that contributes to its development and persistence. Relevant aspects of this complex relationship are detailed here:

Notifications, by offering constant stimuli through messages, alerts, and updates, activate reward centers in the brain similarly to addictive substances. This constant stimulation can create a dopamine-seeking cycle, contributing to the development of addiction.

Digital addiction often manifests in a compulsive ritual of checking notifications. Anticipation of receiving a new alert and the associated reward reinforce the behavior, leading to a continuous need to check the device for digital validation.

The anxiety experienced when separated from the device, known as nomophobia, is exacerbated by the expectation of notifications. Concern about missing relevant information or the constant need to stay connected can intensify digital addiction and emotional dependence on technology.

Digital addiction to notifications can hinder the ability to disconnect. The discomfort or restlessness when away from the screen reinforces the need to stay constantly connected, creating a difficult-to-break vicious circle.

Constant notification interruptions can have a significant impact on productivity. Difficulty maintaining focus and completing tasks without digital distractions contributes to a sense of loss of control, further fueling addiction.

Although notifications are intended to facilitate social connection, digital addiction often results in superficial interactions. The constant search for validation can overshadow the quality of relationships, contributing to a sense of loneliness despite being constantly connected.

Recognizing digital addiction and developing strategies for digital detox are essential steps. Setting screen time limits, practicing regular disconnection, and seeking support to change habits are key approaches to address this issue.

The relationship between notifications and digital addiction underscores the importance of conscious management of online time. Balancing digital connection with periods of healthy disconnection becomes crucial to preserve mental health in the digital age.

Frequent notifications can lead to distraction and social disconnection. Constantly being attached to the phone can affect face-to-face interactions, generating social anxiety and contributing to a sense of emotional disconnection.

The proliferation of notifications on mobile devices has introduced a contemporary phenomenon: digital distraction and social disconnection. These trends, fueled by constant alert interruptions, can have significant effects on the quality of our interactions and personal relationships. Key aspects are analyzed here:

Constant notifications become barriers to mindfulness. Inability to fully engage in a conversation or activity due to digital interruptions generates distraction, weakening the quality of attention and negatively affecting emotional connection with others.

Digital distraction can erode the quality of interpersonal relationships. Lack of complete attention during face-to-face interactions can be perceived as lack of interest or respect, creating tensions and contributing to a sense of emotional disconnection.

Being constantly alert to notifications can generate digital social anxiety. Concerns about online responses, digital validation, and anticipation of new alerts can affect confidence in offline interactions, contributing to social disconnection.

Perpetual digital distraction can also lead to emotional disconnection. Lack of emotional commitment in significant moments due to divided attention between the digital and real worlds can create barriers in relationships, undermining the depth of human connections.

Constant distraction can hinder the ability to empathize. Lack of mindfulness makes it difficult to deeply understand others' emotions and experiences, contributing to a decrease in empathy and emotional connection.

Establishing moments and spaces without digital distractions becomes crucial. Creating notification-free environments allows for richer and more meaningful experiences, promoting mindfulness and genuine connection in social interactions.

Recognizing and rebuilding digital habits is essential to counteract distraction and disconnection. Setting limits on phone use, practicing mindfulness, and fostering awareness of the impact of digital distraction are key steps toward social and emotional reconnection.

Conscious management of notifications and the promotion of balanced technology use are crucial to preserve the quality of our interactions and relationships in a digital era.

Anxiety related to notifications contributes to technological stress. The pressure to keep up with the constant stream of digital information can create additional tensions and negatively impact mental health.

Constant connection to digital devices has brought not only conveniences but also psychological challenges, with technological stress being one of them. This phenomenon arises from anxiety related to notifications and the constant pressure to stay updated with digital information. Key aspects are explored here:

The constant arrival of notifications contributes to information overload. Continuous access to digital information can be overwhelming, generating stress when trying to process and assimilate large amounts of data simultaneously.

The expectation of immediate responses due to notifications creates constant pressure. The feeling of always being available to respond to messages, emails, or updates can generate stress, as there is a perceived need to be constantly connected.

Technological anxiety can also affect productivity. The constant interruption of notifications can fragment the time dedicated to important tasks, reducing efficiency and generating stress related to time management.

The idea of temporarily disconnecting can create anxiety. Dependence on notifications for information and the concern about missing something important can contribute to technological stress when considering the idea of disconnection.

The competition to maintain a positive digital presence also contributes to stress. Concerns about social validation through likes, comments, and followers can create technological anxiety when trying to maintain a flawless digital image.

Effective management of technological stress involves adopting digital self-care strategies. Setting screen time limits, practicing regular disconnection, and prioritizing emotional well-being are crucial approaches to mitigate anxiety associated with notifications.

Creating awareness of the impact of technological stress is fundamental. Understanding how notifications and digital pressure affect mental health can inspire individuals to adopt a more balanced approach to technology.

Effective management of technological stress involves recognizing and addressing tensions associated with constant digital connection. By implementing self-care strategies and fostering a healthy relationship with technology, it is possible to reduce the negative impact of technological stress on mental health.

Developing effective strategies to manage notifications is crucial. Setting time limits, using "Do Not Disturb" modes, and prioritizing important communications are key steps to reduce anxiety associated with notifications.

Defining specific periods to check notifications helps avoid constant interruption. Setting time boundaries allows for dedicated moments during the day to respond to and process digital information.

Using the "Do Not Disturb" function on mobile devices during certain periods, such as work or rest time, helps minimize interruptions and reduces the pressure to respond immediately.

Configuring priority notifications for important messages or key contacts allows for filtering and addressing essential communications first, reducing anxiety related to the constant need for response.

Disabling non-essential notifications contributes to a quieter digital environment. Removing alerts from applications that do not require an immediate response helps reduce anxiety associated with information overload.

Scheduling regular moments of total disconnection is crucial. Establishing periods without notifications, especially before bedtime, allows the mind to relax and reduces anxiety linked to the pressure of always being connected.

Incorporating mindfulness practices helps manage technological anxiety. Being aware of the emotional impact of notifications and practicing mindfulness can improve the relationship with technology.

Communicating to colleagues, friends, and family about specific availability periods helps set clear expectations. This reduces the pressure to respond immediately and fosters a shared understanding of digital limits.

Regularly reviewing notification settings and adjusting them according to changing needs helps maintain a healthy balance. Adapting to circumstances allows for ongoing and effective management of technological stress.

By adopting these strategies, individuals can empower themselves to control digital intrusion, reduce anxiety associated with notifications, and promote a more conscious and balanced use of technology in their daily lives. The role of constant notifications on mobile devices in anxiety and stress is multifaceted.

10.Technology and Personal Relationships

Technology has transformed the way we relate to and communicate with each other, presenting both benefits and challenges to our interpersonal connections.

Technology facilitates instant connection through messages, calls, and social networks. While this allows for easy communication, it can also lead to more superficial and less meaningful interactions.

The digital era has transformed how we connect, offering unprecedented ease in staying in touch with people worldwide. However, this instant connectivity often coexists with the paradox of superficiality in interactions. Let's explore how digital connectivity can impact the depth of our relationships:

The ability to send messages or make calls at any time has fostered a culture of instant gratification. While this facilitates connection, it can also lead to lighter and less committed interactions, prioritizing speed over depth.

Platforms like text messages and social networks often limit the length and complexity of our communications. The brevity of messages can result in more superficial interactions, as details and nuances that would enrich the conversation are omitted.

Digital connectivity may lack the context found in face-to-face interactions. The absence of non-verbal cues and tone of voice can sometimes hinder the accurate interpretation of messages, increasing the risk of misunderstandings.

Although constant connectivity allows us to stay informed, it can also interrupt deep conversations. Notifications and the pressure to respond quickly can distract from meaningful moments.

In an attempt to present a positive online image, people may carefully curate their digital identities. This can lead to a lack of authenticity, as mainly positive aspects of life are shared, contributing to a superficial representation.

Digital connectivity can pose challenges in building meaningful relationships. Online interactions may not offer the same emotional depth as face-to-face experiences, affecting the quality of our connections.

Digital platforms may not be the best environment for addressing complex or emotional issues. The lack of empathy and limitations in emotional expression can hinder effective conflict resolution.

Amidst the ease of digital connection, it's crucial to seek a balance between communication speed and the depth of relationships. Awareness of the need for meaningful and reflective moments can counteract the trend towards superficiality.

Ultimately, digital connectivity offers extraordinary opportunities, but the key lies in using it consciously to nurture meaningful and authentic relationships.

Communication through virtual platforms is often asynchronous, meaning it does not occur in real-time. This can lead to misunderstandings due to the lack of non-verbal signals and tone of voice, impacting the quality of communication.

Asynchronous communication, a distinctive feature of digital platforms, poses unique challenges that influence the quality and depth of our interactions. Let's explore how the lack of temporal synchronization can affect the way we communicate:

Asynchronous communication means responses are not immediate, creating temporal gaps between messages. This disconnection can hinder the development of a natural flow in conversation and affect the perceived urgency in communication.

Non-verbal signals, such as facial expressions and body language, are crucial elements in face-to-face communication. In asynchronous environments, the absence of these signals can lead to misunderstandings, as the interpretation of messages is limited to written content.

The lack of tone of voice in written communication can create ambiguity in message interpretation. The same phrase can be perceived differently based on the tone attributed by the receiver, leading to misunderstandings.

The lack of immediacy in responses can result in impulsive replies. Reflection and careful thought, common in face-to-face communication, may be affected by the temporal distance between messages, contributing to unnecessary conflicts.

Writing may lack the emotiveness present in oral communication. Words may seem colder or more distant, and emotional expressions may be more challenging to convey, affecting emotional connection in interactions.

Given the absence of nuances in in-person communication, there's a greater need to express oneself clearly and precisely in asynchronous environments. The lack of immediate opportunities to clarify can lead to misunderstandings.

Building relationships may be slower in asynchronous communication, as connection doesn't develop in real-time. This can affect the speed at which meaningful relationships are established and the depth of interpersonal connection.

While asynchronous communication offers convenience in many contexts, it's vital to recognize and address these challenges to maximize the effectiveness and authenticity of our digital interactions.

While social media can strengthen connections, it also introduces the risk of constant comparison and the illusion of stronger connections than they really are. Online interactions sometimes do not reflect the complexity of in-person relationships.

The intersection between social media and personal relationships presents a fascinating duality, where digital connections can both strengthen and challenge the nature of our relationships. Let's explore how this duality manifests in the fabric of our lives:

Social media offers a platform to stay connected with friends and family, overcoming geographical barriers. Sharing moments, updates, and messages through these platforms can strengthen bonds, providing a means to be present in each other's lives.

However, exposure to others' lives on social media can trigger constant comparison. The selective nature of what is shared on social networks can create distorted perceptions of others' lives, generating anxiety and unnecessary pressures to meet perceived standards.

Online interaction, though constant, doesn't always reflect the complexity and depth of face-to-face relationships. The illusion of stronger connections than they really are can arise, as the quality of the connection doesn't always directly translate into the digital world.

They may lack nuances present in face-to-face communication. The absence of facial expressions, tone of voice, and body language can make message interpretation more challenging, contributing to misunderstandings.

Social media often fosters an idealized version of life. The pressure to maintain a positive image can inhibit the vulnerability and authenticity crucial for genuine connections. The difficulty in showing less polished sides can affect relationship quality.

While social media facilitates connection with a broad network, there's a risk that these connections may be more superficial compared to deep, meaningful relationships built over time in real life.

The relationship between social media use and mental health is also at stake. Constant exposure to others' lives can contribute to anxiety and low self-esteem, especially when comparing digital life to one's own.

Navigating this duality is crucial, cultivating a healthy balance between digital connectivity and face-to-face relationships, recognizing challenges, and seizing opportunities to build authentic and meaningful connections.

The constant presence of electronic devices during face-to-face interactions can lead to distractions. Lack of mindfulness can affect the quality of communication and diminish intimacy in relationships.

The omnipresence of electronic devices in face-to-face interactions poses a significant challenge to authentic connection and the quality of personal relationships. Let's examine how technological distractions impact the dynamics of our daily interactions:

Notifications, alerts, and sounds from devices can constantly interrupt conversations. This interruption can hinder the development of deep and meaningful dialogue, as people may feel the need to respond quickly to digital distractions.

Lack of mindfulness and constant interruptions can reduce intimacy in relationships. Emotional connection and the depth of conversation may decrease when digital distractions take center stage.

Instead of strengthening social bonds, technological distractions can contribute to disconnection. The inability to maintain a conversation without distractions can lead to a sense of distance and a lack of commitment in relationships.

Prioritizing technology during face-to-face interactions can send a message that the screen is more important than the person present. This perception can negatively affect self-esteem and the quality of the relationship.

Over time, the constant presence of devices can become an ingrained habit. Breaking this habit can be challenging and may require conscious efforts to refocus on human connection without digital distractions.

Constant connection with electronic devices during social interactions can contribute to stress and anxiety. The need to always be connected can create additional pressures and affect mental health.

In addressing technological distractions, it's essential to set clear boundaries and practice mindfulness during face-to-face interactions. Cultivating an environment where relationships are prioritized over digital distractions contributes to building more authentic and meaningful connections.

Virtual communication platforms, while valuable, sometimes cannot fully replicate the richness of face-to-face interactions. The absence of physical contact and limitations in non-verbal signals can affect emotional connection.

Virtual communication, despite its convenience and usefulness, presents specific challenges that can influence the quality and depth of interactions. Let's explore some of these challenges and how they impact communication:

Virtual platforms often lack the richness of non-verbal signals present in face-to-face communication. The absence of facial expressions, gestures, and tone of voice can make the complete interpretation of emotions and nuances in a conversation difficult.

Virtual communication doesn't allow for physical contact, an integral part of human interactions. The lack of hugs, handshakes, or other physical gestures can affect emotional connection and the sense of closeness.

Although emoticons and emojis attempt to compensate for the lack of emotional expression, they cannot always convey the complexity of human emotions. Virtual communication may have limitations in expressing emotions fully.

Virtual interactions may lead to a certain disconnection from reality. Physical separation can make people feel less engaged or less aware of the consequences of their words and actions compared to face-to-face interactions.

The lack of non-verbal signals and the absence of physical contact can increase the risk of misunderstandings in communication. Written words can be interpreted differently, potentially leading to conflicts or confusion.

Extended use of devices for virtual communication can lead to screen fatigue. This not only affects the quality of interaction but can also contribute to exhaustion and digital disconnection.

Virtual communication can make it challenging to read the overall context of a conversation. The absence of ambient and visual cues can make it harder to fully understand the meaning behind words.

Despite these challenges, virtual communication remains a valuable tool. It's important to recognize these limitations and, when possible, complement virtual communication with face-to-face interactions to strengthen emotional connections and promote a more comprehensive understanding.

Online overexposure often translates to a lack of privacy and can affect people's ability to share more intimate aspects of their lives in face-to-face relationships.

The lack of intimacy in the digital age is a growing concern as people share a considerable amount of information online. This phenomenon can have significant impacts on the quality of face-to-face relationships. Let's analyze how online overexposure can affect intimacy:

The presence on social networks and other digital platforms often involves constant exposure to others' lives. The amount of shared information can diminish the sense of privacy and contribute to the feeling of always being under the scrutiny of others.

Online overexposure can affect people's ability to share more intimate aspects of their lives in face-to-face relationships. The feeling that others already know many details can decrease the inclination to share more personal experiences in person.

In a digital environment where sharing personal details has become the norm, some individuals may feel pressured to share more than they are comfortable with. This can affect the authenticity of face-to-face interactions.

Although people share a significant portion of their lives online, they often do so selectively. This careful selection can impact others' perception of a person's authenticity, creating a gap between the digital image and reality.

Lack of privacy can impact trust in interpersonal relationships. People may become more cautious about sharing intimate information, concerned about how it might be perceived or used in the digital space.

Effective management of privacy becomes crucial in a digital world. Properly configuring privacy settings on social networks and being aware of what is shared are essential steps to preserve intimacy.

It is fundamental to create and maintain safe spaces for intimacy in face-to-face relationships. These spaces allow for sharing personal experiences without the concern of constant exposure experienced in the digital realm.

Fostering meaningful and authentic conversations outside the digital sphere contributes to building more intimate and genuine relationships. These face-to-face interactions allow for deeper connection without the pressure of online overexposure.

By balancing engagement in the digital world with the preservation of privacy and intimacy, individuals can cultivate healthier and more authentic relationships both online and offline.

Technology can create expectations of constant availability, testing relationships. The pressure to respond immediately can generate stress and affect the quality of face-to-face interactions.

In the digital era, technology has created expectations of constant availability, which can have a significant impact on interpersonal relationships and face-to-face communication. Let's explore how these expectations can positively and negatively affect the quality of interactions:

Technology allows for instant connection, facilitating real-time communication. This can be beneficial for staying in touch but also creates the expectation of quick responses and constant availability.

The speed of digital communication often generates pressure to respond rapidly to messages, emails, and other forms of communication. This pressure can create stress and affect the quality of responses.

Expectations of constant availability can contribute to stress and anxiety. The need to always be available can lead to concerns about disconnection and negatively impact mental health.

The pressure to respond immediately can affect personal relationships. People may feel frustrated or misunderstood if responses are not instant, leading to unnecessary tensions.

Constant availability through mobile devices can interrupt present moments in face-to-face interactions. The need to respond to notifications can distract and decrease the quality of real-time connection.

It is essential to establish clear boundaries regarding availability. Defining times when it's okay not to respond immediately helps balance digital connection with the need for personal space and mindfulness in face-to-face situations.

Open and transparent communication about availability expectations can prevent misunderstandings. Sharing response preferences and setting boundaries clearly promotes mutual understanding.

Reserving moments for quality time without devices is essential. Establishing technology-free spaces allows people to fully focus on face-to-face interactions without digital distractions.

Educating about the importance of personal spaces and the need to disconnect at certain times can shift cultural expectations and promote a more balanced approach to constant availability.

Prioritizing meaningful interactions over immediate responses helps maintain the quality of relationships. Focusing on the depth of connection rather than the speed of response contributes to more authentic relationships.

By balancing digital connection with personal space and mindfulness in face-to-face interactions, individuals can cultivate healthier and more sustainable relationships in the digital age.

On the other hand, technology enables long-distance relationships, facilitating communication and maintaining meaningful connections despite physical distance.

Technology has transformed the way we maintain and nurture long-distance relationships, acting as a bridge that facilitates meaningful connection and communication. Let's explore how technology has facilitated and enriched these relationships:

Instant Messaging Platforms, Video Calls, and Social Media: These tools allow for instant communication, bridging the geographical gap and creating a sense of closeness despite physical distance.

Video Conferencing: Video conferences have transformed the experience of long-distance relationships. Being able to see each other face to face, share experiences, and engage in activities together virtually helps maintain emotional connection.

Social Media: Social platforms provide a window into the daily lives of distant loved ones. Sharing photos, updates, and participating in online conversations contributes to staying informed about each other's lives.

Collaborative Online Tools: Online collaborative tools make it easy to participate jointly in projects, games, or creative activities. This not only strengthens the connection but also provides shared experiences.

Celebrating Events: Technology allows for the celebration of events from a distance, from virtual birthday parties to participation in special occasions through live streams. This enables sharing meaningful moments despite physical separation.

Group Messaging Apps: Group messaging apps facilitate simultaneous connection with multiple people. This is especially valuable for maintaining cohesion in dispersed groups of friends or families.

Online Gaming: Engaging in online games with distant loved ones offers a fun way to interact and share experiences. Friendly competition and collaboration strengthen bonds.

Traditional Correspondence: Although traditional, the exchange of letters and emails remains an intimate form of communication. Technology facilitates this process, enabling continuous correspondence.

Virtual Events: Participating in virtual events, from conferences to concerts, provides an opportunity to share experiences even when loved ones are geographically distant.

Constant Availability: The constant availability through technology allows for instant emotional support. Being virtually present in difficult moments contributes to ongoing emotional connection.

By leveraging the possibilities offered by technology, long-distance relationships can remain vibrant and meaningful. Technology acts as a bridge that shortens the physical distance, enabling personal connections to thrive despite geographical separation.

Digital communication sometimes lacks the empathy that can be more easily conveyed in face-to-face interactions. The absence of facial expressions and body language can limit emotional understanding.

The evolution of communication toward digital platforms has introduced specific challenges in terms of expressing and receiving empathy. Let's explore how digital communication can affect the ability to express and perceive empathy:

Empathy is communicated not only through words but also through facial expressions, gestures, and tone of voice. In digital environments where these non-verbal signals are limited or nonexistent, conveying empathy can be more challenging.

Written communication, common in digital platforms, can lead to misunderstandings. The absence of intonation and body language can make messages be perceived differently from the original intention, affecting the interpretation of empathy.

Digital conversations sometimes lack the contextual sensitivity found in face-to-face interactions. Empathy thrives on a deep understanding of the situation, and the lack of context can limit the ability to empathize fully.

The instant nature of digital communication can lead to quick and superficial responses. Empathy, which often requires reflection and consideration, may be compromised in an environment where speed is prioritized.

Expressing complex emotions and emotional nuances can be challenging in digital communication. The limitations of the format can make it difficult to convey the full range of emotions that accompany empathy.

The physical separation in digital communication can create a perceptual distance between interlocutors. This distance can affect the perception of empathy, as emotional connection may seem less tangible.

Lack of privacy and overexposure on digital platforms sometimes reduce the sense of intimacy. Empathy, which often flourishes in intimate settings, can be affected by the lack of private spaces.

Although technology has advanced in creating virtual experiences, virtual empathy still faces challenges. Simulating real presence and emotions through screens may not fully replicate face-to-face empathy.

Excessive digital communication can sometimes lead to emotional disconnection. Constant overexposure can desensitize individuals, affecting their ability to connect emotionally and express empathy.

Despite these challenges, it's important to recognize that technology has also created new ways to express empathy, such as online support, participation in virtual support communities, and instant global connection. Adapting and being aware of these dynamics allow mitigating challenges and fostering empathy in the digital world.

Achieving a healthy balance between technology and personal relationships is crucial. Setting boundaries and practicing mindfulness during face-to-face interactions can enhance the quality of interpersonal connections in an increasingly digitized world.

In an increasingly digitized world, finding a healthy balance between technology and personal relationships is essential to preserve authenticity and the quality of our connections. Here are some strategies to achieve this balance:

Define clear limits on the use of technology during face-to-face interactions. Establishing device-free moments allows for a more authentic and meaningful connection.

Mindfulness during personal interactions involves being fully present in the moment. By minimizing digital distractions and focusing on the current conversation or activity, a deeper connection is promoted.

Allocate specific spaces for technology use to help separate digital interactions from personal ones. This contributes to maintaining attention on relationships when it's most relevant.

While technology facilitates long-distance communication, encouraging face-to-face interactions strengthens emotional connections. In-person conversations allow for the complete expression of emotions and empathy.

Instead of solely focusing on electronic devices, engaging in joint activities strengthens personal bonds. Going out together, participating in outdoor activities, or sharing hobbies creates meaningful memories.

In a hyperconnected digital world, prioritizing the quality of interactions over quantity is key. Rather than seeking constant online validation, focus on building deep and meaningful relationships for emotional well-being.

Learning to appreciate moments of silence in each other's company is important. Not every moment needs to be filled with conversations or electronic devices. Sometimes, connection is found in shared tranquility.

Fostering relationships outside digital platforms provides shared experiences not conditioned by the constant pursuit of exciting online events. These relationships anchor connection in tangible reality.

Reflecting on the need for constant online validation is crucial. Redefining personal value away from digital popularity indicators contributes to a healthier self-image.

Designating spaces at home or during social activities where technology is not present allows for disconnecting and enjoying full presence in the current moment.

By incorporating these strategies, a healthy balance can be achieved, leveraging the benefits of technology without compromising the authenticity and depth of personal relationships.

11.Impact of Blue Light on Sleep

Exposure to blue light emitted by electronic screens, such as smartphones, tablets, and computers, can have a significant impact on our sleep cycle and the quality of rest. Here, we explore the effects of blue light and some strategies to mitigate its impact:

Blue light is a short-wavelength, high-energy light commonly found in electronic devices. This light is beneficial during the day as it enhances alertness and performance, but its nighttime exposure can suppress melatonin, the hormone that regulates sleep.

Blue light, being a short-wavelength and high-energy light, is part of the visible light spectrum that influences sleep regulation. During the day, exposure to blue light is beneficial because it helps keep us alert and improves cognitive performance. However, at night, continuous exposure to blue light can interfere with melatonin production, negatively affecting our sleep.

Melatonin is a hormone that regulates the sleep-wake cycle, and its production is influenced by ambient light. The presence of light, especially blue light, signals to the brain that it is daytime, inhibiting the release of melatonin and complicating the process of falling asleep.

For this reason, reducing exposure to blue light in the hours before sleep, especially from electronic device screens, can be an effective strategy to improve sleep quality. Using blue light filters, activating night modes on devices, and limiting screen use before bedtime are recommended practices to care for our sleep health in a digital environment.

Exposure to blue light, especially in the hours before sleep, can interfere with melatonin production, making it difficult to fall asleep. This can lead to disruptions in the circadian rhythm and affect the quality of rest.

Exposure to blue light in the hours before sleep can suppress melatonin production, disrupting the circadian rhythm, which is the internal biological clock that regulates sleep-wake cycles.

Melatonin is essential for initiating and maintaining sleep. When exposed to blue light, especially from electronic devices such as phones, tablets, and computers, the body interprets this light as a signal that it is daytime and, consequently, reduces melatonin production. This interference with the natural regulation of sleep can result in difficulties falling asleep, insomnia, and generally poorer sleep quality.

Adopting practices that reduce exposure to blue light before bedtime, such as activating night modes on electronic devices or using blue light filters, can be beneficial in preserving sleep quality and maintaining a healthy circadian rhythm.

Blue light can deceive the brain into thinking it's daytime, even in dark environments. This can disrupt the natural sleep cycle, making it harder to fall asleep and affecting the duration and quality of rest.

The ability of blue light to deceive the brain and signal that it's daytime, even in darkness, can have a significant impact on the natural regulation of sleep. This deception can affect the ability to fall asleep and the quality of rest.

It is important to be aware of how exposure to blue light, especially in the hours before bedtime, can influence our circadian rhythm and, consequently, our sleep patterns. Adopting habits that minimize exposure to blue light, such as reducing the use of electronic devices before bedtime or using blue light filters, can contribute to maintaining a healthier sleep cycle.

Constant exposure to blue light before bedtime has been associated with less restorative sleep. People who use electronic devices before sleep tend to experience a decrease in sleep quality and may wake up feeling less rested.

Sleep quality can be negatively affected by constant exposure to blue light, especially before bedtime. The use of electronic devices at night can interfere with the body's ability to produce melatonin, the sleep hormone, resulting in less restorative sleep and a feeling of fatigue upon waking.

It is crucial to be aware of these effects and consider practices that minimize exposure to blue light before sleep, such as establishing periods of digital disconnection, using night mode settings on devices, and creating a sleep-friendly environment in the bedroom. These measures can contribute to improving sleep quality and promoting healthier rest.

Using apps or settings on devices that reduce blue light emission. Activating night modes on devices that automatically adjust the color temperature towards warmer tones in the evening. Using warm ambient lights instead of bright lights at home before bedtime. Avoiding the use of electronic devices at least an hour before bedtime to allow melatonin to be naturally released. Maintaining a regular bedtime and wake-up routine helps synchronize the circadian rhythm. Keeping the bedroom dark and cool, and limiting screen exposure at night. Recognizing the impact of blue light on sleep is the first step in taking proactive measures. With awareness and adjustments to our nighttime habits, we can better care for our sleep health in the digital age.

Awareness of the impact of blue light on sleep is essential to take actions that promote healthy rest. By understanding how exposure to blue light can affect melatonin production and alter the sleep cycle, individuals can adopt practices that minimize these negative effects.

Implementing changes in nighttime habits, such as reducing the use of electronic devices before bedtime and adjusting the lighting in the environment, can make a significant difference in sleep quality. By prioritizing sleep self-care, there is a significant contribution to overall health and well-being.

By adopting these strategies and being mindful of our exposure to blue light, we can significantly improve the quality of our sleep and promote restful sleep that contributes to our overall well-being.

Developing a calm bedtime routine can prepare the mind and body for sleep. This could include activities such as reading a book, taking a warm bath, or practicing meditation.

Ensuring that the room is dark, quiet, and cool can enhance conditions for falling asleep. Blackout curtains and the use of earplugs may be helpful.

Avoiding caffeine and heavy meals before bedtime can help prevent sleep disruptions. Opting for light snacks and avoiding caffeine at least several hours before bedtime is beneficial.

Going to bed and waking up at the same time every day, even on weekends, can help regulate the internal biological clock and improve sleep consistency.

Practicing mindfulness can reduce stress and anxiety, contributing to a more relaxed mental state before bedtime.

Regular physical activity is linked to better sleep quality. However, it is advisable to avoid intense exercise right before bedtime.

Prioritizing these habits can make a significant difference in sleep quality and, consequently, overall well-being.

12.Authenticity on Social Media

Exploring the theme of authenticity on social media is crucial given the growing gap between online representation and reality.

Social media is often used as a platform to share the highlights and positive moments of life. This selectivity can create a gap between how a person presents themselves online and their everyday reality, leading to a distorted representation.

Editing tools and filters on social media allow users to enhance and modify their photos before sharing them. This can lead to an idealized representation that doesn't fully reflect reality.

People tend to share the most significant and positive moments on their profiles, creating a narrative that may deviate from the complexity of everyday life. This can generate a biased image of reality.

There is cultural pressure to maintain a positive online presence. Users may feel compelled to highlight the good aspects of their lives, contributing to a partial and often unrealistic representation.

Viewers of these representations may fall into the trap of constant comparison. By seeing others' seemingly perfect lives, it's easy to feel dissatisfied with one's own reality, leading to anxiety and low self-esteem.

The disconnect between reality and online representation can be amplified by a group effect, where people feel pressured to follow the trend of sharing only the positive aspects of their lives.

The lack of authenticity in online representation can have a negative impact on mental health. The pressure to maintain a positive image and constant comparison can contribute to issues like anxiety and depression.

Fostering a culture of authenticity on social media involves recognizing and celebrating the complexity of real life. Promoting honesty and openness can counteract the negative effects of the gap between reality and online representation.

Addressing this discrepancy requires a collective effort to change cultural expectations and foster a digital space where authenticity is valued and appreciated.

Constant comparison with others' seemingly perfect lives on social media can negatively impact self-esteem. Users may feel pressured to meet unrealistic standards, leading to feelings of inadequacy and lack of worth.

Social media often displays a selected and edited version of someone's life. This can create the illusion that everyone else has perfect lives, exacerbating feelings of inadequacy.

The pursuit of validation through likes, comments, and followers can become a source of self-esteem for some. Lack of interaction or comparison with those receiving more attention can lead to insecurity.

The nature of social media can encourage a constant cycle of comparison. People may see themselves in an implicit competition for online attention and approval.

Social media often presents a positive and optimistic version of people's lives. This positive bias can make others feel that their own lives don't measure up, negatively affecting their self-image.

The connection between constant comparison on social media and mental health is significant. Anxiety, depression, and low self-esteem are possible consequences of continuous exposure to idealized representations of others' lives.

Promoting a shift in the focus of social media, moving away from comparison and emphasizing genuine connection, can contribute to an improvement in self-esteem. Valuing authentic experiences over idealized representations can be crucial.

Addressing these issues involves an effort at both individual and cultural levels to promote authenticity, compassion, and a more realistic understanding of life on digital platforms.

The quest for validation in the form of likes, comments, and followers can drive behaviors seeking approval rather than authenticity. Digital interactions often become a constant pursuit of external validation, which can be detrimental to intrinsic self-esteem.

Indeed, the culture of digital validation can have a significant impact on how people perceive their own value and self-esteem. Here are some additional aspects that delve into this topic:

Validation in the form of likes and comments triggers the release of dopamine in the brain, creating a sense of reward. This mechanism can lead to a constant quest for more validation, contributing to approval-driven behaviors.

Sometimes, people may sacrifice authenticity in pursuit of popularity. Creating content or presenting life in a way that attracts more likes can lead to a distorted representation and loss of authenticity.

When self-esteem is directly tied to the quantity of likes or online interactions, it can become fragile. Fluctuations in digital attention can negatively impact emotional well-being.

Digital validation can also foster constant comparison. People may gauge their worth based on online popularity, comparing themselves to others and feeling dissatisfied if they perceive they are not on par.

Excessive dependence on online validation can lead to emotional disconnection from reality. Constant attention to the screen can distance individuals from meaningful interpersonal relationships outside the digital realm.

Fostering a culture that values authenticity over popularity can counteract some of the negative impacts. Promoting genuine self-expression and supporting the diversity of experiences can contribute to a healthier online environment.

It is essential to raise awareness about these aspects and foster a culture that celebrates authenticity.

The pressure to maintain a positive image online can generate additional stress. Individuals may feel the need to conceal less positive aspects of their lives, contributing to a lack of authenticity in online representation.

The pressure to maintain a positive image online can have a significant impact on emotional well-being and the authenticity of digital interactions. Some additional aspects to consider include:

Concern for others' judgment can lead to self-censorship. People may avoid sharing aspects of their lives they perceive as less positive, contributing to the creation of a biased online narrative toward the positive.

The pressure to maintain a positive image can intensify the comparison cycle. By seeing others' seemingly perfect lives, individuals may feel the need to match or surpass those representations, contributing to a sense of online competition.

The expectation of maintaining a positive image can sometimes hinder genuine support-seeking. People may fear showing vulnerability or weaknesses, limiting opportunities for real connection and emotional support.

The constant representation of a positive life can create dissonance between reality and online perception. This mismatch can contribute to stress, anxiety, and other mental health issues while trying to maintain a constant facade.

Persisting in presenting a positive image can have long-term consequences on mental health. Lack of authenticity and constant pressure can contribute to emotional exhaustion and a sense of disconnection from one's identity.

Promoting the importance of authenticity and acceptance of human complexity online can help alleviate the pressure to maintain a positive image. Celebrating authenticity contributes to building more understanding and supportive online communities.

In addressing these challenges, it is crucial to create online spaces where people feel comfortable sharing both the positive and challenging aspects of their lives, thus fostering more authentic and meaningful connections.

Fostering self-acceptance and understanding that real life is inherently imperfect can help counteract the negative effects of the gap between online representation and reality.

The promotion of self-acceptance is essential in a digital world where constant comparison can negatively impact self-esteem and emotional well-being. Here are some additional points about the importance of fostering self-acceptance:

Self-acceptance contributes to emotional well-being by allowing people to embrace themselves with all their imperfections. This reduces self-criticism and fosters a more compassionate attitude toward oneself.

By promoting the idea that it's okay to be oneself, the need for constant comparison with others is reduced. Personal acceptance decreases the pressure to meet unrealistic standards.

Self-acceptance is a key component of emotional resilience. People who accept themselves are better equipped to cope with challenges and overcome obstacles without being overwhelmed by self-criticism.

Self-acceptance also contributes to healthier interpersonal relationships. When individuals accept themselves, they are more likely to establish relationships based on authenticity and honesty.

Self-acceptance empowers individuals by acknowledging and embracing their uniqueness. This can lead to increased self-confidence and a willingness to approach challenges and opportunities with a positive mindset.

Promoting self-acceptance contributes to the creation of a more inclusive online culture. When people feel accepted and valued regardless of their differences, a more respectful and tolerant online environment is fostered.

The promotion of self-acceptance not only benefits individuals on an individual level but also contributes to the creation of healthier and more compassionate online communities.

Ultimately, cultivating a culture of authenticity on social media is essential for building more genuine connections and promoting mental health in an increasingly influential digital environment.

13.Effect of Image Filter

The widespread use of image filters in photos shared on social media has sparked discussions about how it affects the perception of beauty and self-image. Image filters can significantly alter a person's appearance, smoothing the skin, adjusting facial contours, and changing lighting. This creates an idealized version that can distort the perception of reality and generate unrealistic expectations about physical appearance.

The altered perception of reality due to the use of image filters is a relevant phenomenon in the digital era. These filters, often available on social media apps and platforms, provide the ability to modify the appearance of photos in various ways. Some of the most common effects include skin smoothing, facial feature refinement, color changes, and improved lighting. These adjustments can create an idealized and stylized version of the person in the photograph.

The alteration of reality through filters presents several impacts and considerations:

Filtered images can set unrealistic expectations about physical appearance. Those who regularly consume edited images may develop unattainable standards for themselves and others, contributing to social pressure and personal dissatisfaction.

The widespread presence of filtered images on social media can foster a culture of constant comparison. People may compare their unedited appearances with retouched versions, leading to feelings of insecurity and affecting self-esteem.

Those who regularly use filters may experience an impact on their self-image. The gap between the online image and reality can lead to dissatisfaction with natural appearance and contribute to the perception that authenticity is not valued.

The prevalence of filters can emphasize the importance of superficial aesthetics over authenticity. This can have implications for how people value beauty and relate to each other, prioritizing visual perfection over diversity and authenticity.

The growing public awareness of image editing and filter use has led to ethical debates. There is a discussion about whether highly edited images present a fair and honest representation in the digital environment.

Some people advocate for greater transparency in filter use, encouraging users to share images that more accurately reflect their real appearance. This promotion of authenticity contributes to building a more transparent and compassionate digital culture.

Ultimately, the alteration of reality perception through filters highlights the importance of promoting authenticity and self-acceptance in a digital environment that often presents idealized versions of reality.

The frequent use of filters can contribute to social pressure to meet unrealistic beauty standards. Individuals who regularly consume filtered images may feel the need to conform to these expectations, negatively affecting self-image and self-esteem.

The pressure to meet unrealistic beauty standards, amplified by the frequent use of filters, is a significant dynamic in the digital era. This pressure can manifest in various ways and impact mental and emotional health in several ways:

The prevalence of filtered images on social media can lead to constant comparison between natural appearance and edited versions. This can generate dissatisfaction and fuel the perception that one does not meet prevailing aesthetic standards.

People may feel the need to apply filters regularly before sharing their own photos online. This can become a habitual practice to conform to digital beauty expectations, adding an additional layer of pressure and self-demand.

The discrepancy between real appearance and the stylized version through filters can affect self-esteem. Individuals may develop distorted perceptions of their own beauty, contributing to feelings of inadequacy and negative self-evaluation.

The pressure to meet digital standards can influence decisions related to physical appearance. This could include decisions about cosmetic procedures, choice of makeup, and other behaviors aimed at conforming to perceived norms.

The repeated use of filters to meet digital standards can lead to a dependence on these tools for self-esteem. People may start feeling uncomfortable or insecure without the application of filters, affecting their perception of their own beauty without these digital tools.

The growing awareness of the negative effects of unrealistic beauty standards has led to a call for authenticity. Some people seek to counteract these effects by promoting more authentic images and representations of beauty, encouraging personal acceptance and diversity.

The pressure to meet unrealistic standards underscores the importance of fostering a digital culture that values diversity, authenticity, and self-acceptance. By being aware of these challenges, we can work towards building more inclusive and positive online environments.

The presence of filtered images on social media can foster constant comparison. People may compare their unfiltered appearance with retouched images, leading to feelings of inadequacy and fueling a culture of comparison.

Constant comparison is a significant dynamic arising from the prevalence of filtered images on social media. This trend can have a profound impact on the mental and emotional health of individuals:

By comparing their unfiltered photos with retouched images of others, individuals may experience feelings of inadequacy. The discrepancy between reality and digital representation can create the perception that one does not meet the aesthetic standards promoted online.

Constant exposure to enhanced images can create significant pressure for individuals to conform to those standards. This can lead to behaviors such as excessive filter use and the adoption of practices to alter physical appearance to meet unrealistic expectations.

Constant comparison can undermine self-esteem. People may feel they are not attractive enough or do not meet widely accepted beauty standards. This can contribute to insecurity and negatively impact self-perception.

A digital culture that fosters constant comparison may have broader ramifications in society. It can contribute to the creation of an environment where self-evaluation is based on superficial standards and physical appearance rather than deeper aspects of identity and personal worth.

Constant comparison can have an impact on overall emotional well-being. It can contribute to anxiety, depression, and other mental health issues by generating a distorted perception of reality and self-image.

To counteract these effects, it is crucial to promote self-acceptance, diversity, and authenticity online. By fostering a digital culture that celebrates uniqueness and values authenticity, we can contribute to more positive environments.

Excessive use of filters can hinder self-acceptance. People may become more critical of their natural appearance when comparing it to filtered versions, negatively affecting their relationship with their own bodies.

Filters often alter the body's appearance, smoothing contours, refining features, and adjusting proportions. This distortion can lead to an unrealistic perception of how the body should look, contributing to a distorted body image.

Frequent use of filters can make individuals compare themselves to unrealistic aesthetic standards. This can create a sense that natural appearance is not sufficient or that certain beauty ideals exist only in the digital realm.

Those who receive validation and social approval for their filtered photos may feel additional pressure to maintain that image. This can lead to discomfort with the unfiltered appearance and contribute to a constant pursuit of digital perfection.

Constant concern for appearance and self-evaluation based on unattainable standards can have consequences for mental health. It can contribute to anxiety, depression, and disorders related to body image.

The lack of authenticity in presenting a filtered version can affect genuine connection with others. Authenticity and self-acceptance are crucial for building meaningful relationships, and excessive filter use can hinder this process.

Promoting self-acceptance involves fostering a culture that celebrates the diversity of bodies and appearances. Additionally, emphasizing the importance of authenticity in digital interactions can contribute to healthier online environments.

The prevalence of filtered images can create unrealistic expectations in face-to-face interactions. People may expect others to look a certain way, which can affect personal relationships and the perception of authentic beauty.

The prevalence of filtered images can influence the formation of unrealistic expectations about people's appearance in face-to-face situations. This can manifest in various ways:

When people are accustomed to seeing filtered images online, they may develop unrealistic expectations about how individuals should look in everyday life. This can lead to a distorted perception of beauty and generate unnecessary pressures to meet unrealistic standards.

Those who consistently consume filtered content may experience a disconnection between reality and expectations. This can affect their self-image, as they may feel they do not meet supposed beauty norms, even if they are healthy and have a completely normal appearance.

The pressure to meet expectations generated by filtered images can lead individuals to attempt to replicate that appearance in real life. This can create anxiety and affect self-esteem when they perceive they cannot reach those filtered standards.

The discrepancy between expectations based on filtered images and reality can affect interpersonal relationships. People may feel disappointed or dissatisfied when someone's appearance does not align with the perfected images they have seen online.

Promoting the importance of authenticity and acceptance in interpersonal relationships can counteract these unrealistic expectations. Celebrating diversity and recognizing authentic beauty in all its forms contributes to creating healthier and more understanding environments.

Emphasizing the use of filters can place more focus on superficial aesthetics than authenticity. This can affect how people value beauty, prioritizing visual perfection over diversity and authenticity.

The emphasis on filter use can contribute to a culture that values superficial aesthetics over authenticity. Here are some key points related to this phenomenon:

The prevalence of filtered images can distort the perception of reality by presenting idealized versions of individuals. This can lead to the belief that beauty is intrinsically linked to visual perfection, creating a gap between reality and expectations.

When filtered aesthetics become the norm, it can create significant pressure on individuals to meet beauty standards that may be virtually unattainable. This contributes to a culture of constant comparison and the pursuit of unreal perfection.

Overexposure to filtered images can diminish the celebration of diversity in terms of appearance. Authentic beauty, encompassing a wide range of features and bodies, may be sidelined in favor of visual standardization driven by filters.

A culture favoring aesthetics over authenticity can negatively impact self-acceptance. People may feel the need to conform to unrealistic standards, influencing their self-perception and self-esteem.

Promoting the importance of authenticity and celebrating diversity in appearance is essential. This involves recognizing and valuing the unique beauty of each individual, regardless of whether they meet certain filtered visual standards. Authenticity and diversity should be celebrated as fundamental elements of true beauty.

Public awareness of filter use has led to debates about the ethics of image editing and the importance of transparency in presenting images online. Some people advocate for a more authentic representation on social media.

Image editing, especially when used to significantly alter a person's appearance, has sparked ethical debates. There is questioning whether these practices contribute to unrealistic standards and pressure individuals to meet unattainable expectations.

Transparency in image presentation has become crucial. Some people advocate for disclosing whether an image has been edited, providing viewers with the necessary information to interpret it accurately. Honesty in online representation is fundamental to building authentic digital relationships.

There have been movements and campaigns promoting authenticity on social media. Some advocates urge sharing unfiltered images to challenge traditional beauty norms and foster a culture that celebrates diversity and authentic reality.

Public awareness of image editing has also highlighted the impact it can have on the self-esteem of those consuming online content. By understanding that edited images do not necessarily represent reality, people may be more inclined to cultivate a healthy and realistic self-image.

Promoting digital education about image editing and the importance of transparency is essential. Helping people understand how filters and editing tools are used can empower them to critically interpret the images they encounter online.

In summary, awareness of edited reality has prompted a call for transparency and authenticity in online representations, recognizing the importance of building a more honest and healthy digital culture.

Image editing, especially when used to significantly alter a person's appearance, has sparked ethical debates. There is questioning whether these practices contribute to unrealistic standards and pressure individuals to meet unattainable expectations.

The main concern lies in how image editing can contribute to perpetuating unrealistic beauty standards. Image manipulation can create idealized representations that are difficult to attain, affecting people's self-esteem and contributing to social pressure.

There is concern that image editing creates undue pressure on individuals to meet unrealistic expectations. When edited images dominate social media platforms, an environment can be created where authenticity and diversity are overshadowed by a relentless pursuit of visual perfection.

It has been argued that constant exposure to edited images can have a negative impact on mental health, contributing to issues such as low self-esteem, body anxiety, and dissatisfaction with personal appearance.

The debate also addresses the responsibility of media and online platforms. Some voices argue that there should be greater regulation and transparency around image editing, with stricter requirements to disclose when an image has been significantly altered.

On the other hand, there is a growing movement advocating for the promotion of authenticity online. This involves sharing unfiltered or minimally edited images, fostering a culture that celebrates diversity and values reality over artificial perfection.

Some proposed solutions include empowerment through education. Teaching people to understand and question the images they consume can be a powerful tool to counteract the negative effects of edited representations.

Overall, the ethical debate on image editing highlights the need to balance creative and artistic expression with the responsibility to promote a healthy and realistic body image in the digital space.

14. Technology and Self-Control

The relationship between technology and self-control is a crucial topic in the digital era.

Applications and digital platforms are often designed with features that aim to keep users engaged for as long as possible. This can include constant notifications, rewards, and other mechanisms that may hinder self-control and disconnection.

The design of digital applications and platforms plays a crucial role in influencing user behavior and can have significant effects on self-control. Some key aspects of design that can contribute to addiction and hinder self-control include:

Instant notifications, especially those designed to be irresistible, can keep users constantly attentive and contribute to a sense of urgency that makes it challenging to ignore or postpone a response.

The use of rewards, such as pleasant sounds, achievement icons, and positive messages, can activate the brain's reward system, generating a sense of pleasure and encouraging continuous use.

Incorporating game elements, such as points, levels, and competitions, can make using the application more engaging and addictive. Users may feel motivated to spend more time to achieve virtual goals.

Platforms like social networks, video streaming, and news can offer an endless flow of content, keeping users scrolling indefinitely. This can make it difficult to set time limits.

Algorithms that personalize content based on user preferences can increase relevance but may also create a digital "bubble" that reinforces existing opinions and keeps people on the platform for longer.

The constant accessibility from mobile devices facilitates impulsive and frequent use of applications, contributing to the difficulty of disconnecting.

Understanding how these elements influence user behavior is crucial for promoting ethical design practices and helping users develop greater self-control in their relationship with technology. Awareness of intentional manipulation in design can empower individuals to make informed decisions about their digital use.

Social media and other applications can consume significant amounts of time, impacting the balance between online and offline life. The constant updating of feeds and instant availability of content can contribute to a loss of self-control in terms of screen time.

Social media is designed to offer new and updated content constantly. This dynamic nature may make users feel compelled to check and update their feeds frequently to avoid missing relevant information.

Social media notifications can be irresistible, prompting an almost automatic response. The perceived need to respond to notifications can lead to more frequent app usage and, consequently, increased screen time.

Social media platforms often employ design techniques that seek constant attention and interaction. This may include infinite scrolling, personalized suggestions, and other features that keep users engaged.

The nature of social media, where people share highlights of their lives, can foster constant social comparison. This phenomenon may lead to prolonged use as users seek validation and compare their lives to others.

To counter these effects and maintain greater control over screen time, users can consider:

Using built-in features in devices or applications that allow setting daily limits for the use of specific applications.

Allocating specific times of the day to completely disconnect from social media and other applications.

Customizing notification settings to reduce distractions and avoid the need to check constantly.

Monitoring and being aware of the time spent on devices and setting realistic goals to gradually reduce it.

These practices can help balance technology use and preserve time for offline activities and personal relationships.

Constant notifications can disrupt workflow and distract people from their tasks. Lack of self-control in responding immediately to each notification can impact productivity and concentration.

Notifications can interrupt workflow and concentration on a specific task. Each interruption requires time to shift attention and refocus on the original task, affecting efficiency.

Activate "Do Not Disturb" modes during periods of intense concentration or set specific times to check notifications.

The expectation to respond immediately to notifications can create a constant sense of urgency, affecting work quality and increasing stress.

Establish designated periods to review and respond to notifications instead of doing so immediately. This helps maintain a more continuous focus on important tasks.

Frequent notifications can tempt people to engage in multitasking, which generally reduces efficiency and the quality of work.

Prioritize tasks and focus on one activity at a time. Multitasking can decrease the quality of work and increase the risk of errors.

The need to be aware of every notification can create constant distraction, especially in work or academic environments.

Set time limits for device use and notifications. This helps create dedicated periods for concentration without interruptions.

The pressure to respond quickly and stay on top of all notifications can contribute to stress and anxiety.

Set realistic expectations and recognize that not all notifications require an immediate response. Prioritizing mental health is crucial.

By implementing these strategies, individuals can enhance their ability to maintain self-control over notifications, reduce distractions, and improve the quality of their work and personal time.

Exposure to screens before bedtime, whether through phones, tablets, or other devices, can affect the circadian rhythm and hinder sleep. Exercising self-control by setting limits on nighttime device use is essential for sleep health.

Nighttime use of electronic devices, such as phones, tablets, or computers, can have a negative impact on sleep quality due to exposure to the blue light emitted by these screens. Here are some considerations and strategies to exercise self-control and improve sleep health:

Blue light from screens can suppress the production of melatonin, the sleep hormone. This can interfere with the ability to fall asleep and affect the quality of rest.

Establish a "cut-off time" for the use of electronic devices at least an hour before bedtime. This helps natural melatonin production.

Many devices have night modes or blue light filter settings that reduce blue light emission during nighttime hours.

Activate night modes or use applications and settings that decrease the amount of blue light emitted by the screen.

Creating regular sleep routines can help the body recognize when it's time to prepare for sleep.

Set consistent bedtime and wake-up schedules. This helps regulate the circadian rhythm.

Associating the bed with relaxing activities, rather than stimulating ones like device use, can improve sleep quality.

Avoid using electronic devices in bed and reserve that space for sleeping and relaxing.

Instead of reading on electronic devices before bedtime, opting for physical books can be a more relaxing alternative.

Incorporate physical reading into the nighttime routine before sleep.

Exercising self-control regarding electronic device use before bedtime is crucial for promoting healthy sleep habits and contributing to overall well-being.

While some applications are designed to increase productivity, others may contribute to procrastination. Self-control is challenged when the temptation to use less productive applications is high.

Solution: Recognize and use applications designed to increase productivity. Time management apps, to-do lists, and collaboration tools are examples of tools that can be beneficial.

Define clear goals and priorities before using applications. This helps maintain focus on important tasks.

Set time limits for using apps that tend to be distracting or promote procrastination. This may include social media apps, games, and entertainment.

Use focus modes or apps that temporarily block access to distracting apps during specific work periods.

Implement time management techniques, such as the Pomodoro Technique, to divide work into specific time intervals, alternating with short breaks.

Conduct regular self-assessments to identify usage patterns of apps and adjust focus as needed.

Prioritize important and urgent tasks before diving into app usage. This helps focus on what truly matters.

Schedule moments of disconnection, especially during crucial activities or when maximum concentration is needed.

Establish positive rewards for after completing important tasks. This can motivate maintaining focus and avoiding procrastination.

The key is to find a healthy balance between using productive apps and avoiding those that may be distracting or contribute to procrastination. Maintaining a conscious and proactive approach can significantly improve personal productivity.

Some people use apps to monitor and limit their screen time. This approach can help strengthen self-control by providing greater awareness of the time spent on electronic devices.

Screen time monitoring gives individuals a clear view of how much time they spend on electronic devices. This awareness can be the first step in identifying behavior patterns.

By understanding how much time is dedicated to certain apps or online activities, people can set realistic goals to reduce screen time and improve productivity.

Identifying time spent on less productive activities allows individuals to adjust their focus and time toward more meaningful and prioritized activities.

Screen time monitoring facilitates time management by helping people identify where most of their time goes and how they can redistribute it more effectively.

Limiting screen time before bedtime, based on monitoring, can improve sleep quality by reducing exposure to the blue light emitted by screens.

Choosing reliable and secure screen time monitoring apps is crucial. Several apps are available that offer tracking and time limit features.

Customizing notification and time limit settings in monitoring apps can cater to individual needs and help maintain a healthy balance.

Regular self-assessments are important to adjust goals and time limits as personal needs and priorities evolve.

If it's a family or group effort, open communication about goals and time limits can encourage a supportive environment and accountability.

While screen time monitoring is valuable, it's also essential to balance it with periods of complete disconnection to promote overall well-being.

By implementing screen time monitoring consciously and proactively, individuals can take meaningful steps to strengthen their self-control and foster healthier digital habits.

Encouraging the development of healthy digital habits involves setting conscious limits on screen time, scheduling periods of disconnection, and promoting regular self-assessment of technology use.

The development of healthy digital habits is essential to maintaining a positive balance between online and offline life. Here are some key strategies to encourage healthy digital habits:

Define clear limits for daily screen time. This may include specific times of the day dedicated to online activities and designated moments for disconnection.

Plan regular periods of complete disconnection. Establish tech-free days or periods during the day when the phone is turned off and social media is disconnected to promote renewal and reduce dependence.

Integrate mindfulness practices into digital life. This involves being aware and intentional when using devices, focusing on one task at a time, and avoiding multitasking to improve the quality of online connection.

Create balanced routines that include time for offline activities such as exercise, reading, and socialization. Maintaining a balance between online and offline activities contributes to a healthier lifestyle.

Reduce unnecessary notifications to minimize interruptions. This helps maintain focus and prevents constant distraction caused by device alerts.

Set specific goals for technology use. This could include reducing time on social media, limiting online news consumption, or establishing specific times for checking emails.

Encourage outdoor activities. and time in nature. Digital disconnection to connect with the natural environment can have significant benefits for mental and emotional well-being.

Regularly reflect on digital habits and conduct self-assessments. Ask how one feels about screen time and whether current habits align with personal goals and values.

Provide education on the healthy use of technology, especially in family settings. This includes establishing guidelines for children and teenagers and fostering open communication about the importance of a proper balance.

Share goals and challenges with friends, family, or colleagues to provide support and accountability. Working together to develop healthy digital habits strengthens commitment and motivation.

Fostering these habits contributes not only to a more balanced relationship with technology but also to overall well-being and an improved quality of life.

Education on the importance of digital self-control and strategies to maintain a healthy balance between technology and other activities is crucial. This may include promoting digital breaks, offline activities, and practicing mindfulness.

Education on digital self-control is crucial to empower individuals with the necessary skills to effectively manage their relationship with technology. Here are some key aspects that could be addressed in digital self-control education:

Encourage self-assessment and awareness of digital habits. Help individuals recognize behavior patterns, such as excessive device use or dependency on social media.

Inform about the risks associated with excessive technology use, such as visual fatigue, decreased sleep, increased stress, and reduced productivity. Emphasize the importance of maintaining a healthy balance.

Teach practical strategies for setting limits on screen time. This may include using device time limit features, establishing specific schedules for online activities, and the importance of disconnecting at key moments.

Introduce digital mindfulness practices to enhance concentration and reduce multitasking. Teach individuals to be present in the moment and avoid unnecessary distractions when using devices.

Highlight the significance of taking regular digital breaks. Show how short breaks can improve productivity, reduce stress, and foster connection with the surrounding environment.

Present healthy alternatives to digital activities, such as reading physical books, engaging in outdoor activities, exercising, and face-to-face socialization. Demonstrate how these activities can enrich life beyond the screen.

Provide guidance on managing notifications to avoid constant interruptions. Teach individuals to prioritize important notifications and disable those that do not contribute to overall well-being.

Develop self-regulation skills by focusing on goals and priorities. Help individuals identify personal goals and utilize technology as a tool to support those goals rather than becoming a distraction.

Promote a positive attitude toward technology use. Illustrate how technology can be a valuable tool when used consciously and in balance, highlighting examples of positive impact in daily life.

Digital self-control education can help individuals make informed decisions, set healthy limits, and cultivate a balanced relationship with technology in the digital age.

Self-control in the digital era involves being aware of how applications and devices influence our behavior and intentionally setting boundaries to ensure a healthy and balanced use of technology.

15.Technological Stress

Technological stress is a reality in today's society, where dependence on technology and constant connectivity can lead to emotional and physical tensions. The omnipresence of devices and continuous internet connection can create a sense of always being available, making it challenging to disconnect and rest.

The onslaught of information we are constantly exposed to can be overwhelming, causing mental fatigue and difficulty concentrating on important tasks. The culture of quick responses in digital communications can generate anxiety, as instant replies are expected, increasing pressure on individuals.

Social media can contribute to constant comparison with others, leading to anxiety and diminished self-esteem. Concerns about online security and privacy can cause stress, especially when sharing personal information on digital platforms.

Excessive use of electronic devices and social media can lead to addiction, negatively affecting mental health and interpersonal relationships. Defining specific schedules for technology use and setting clear limits on online availability can help reduce stress.

Incorporating regular periods of complete disconnection, such as during meals or before bedtime, can reduce digital fatigue. Implementing strategies to manage information overload, such as prioritizing tasks, using organizational tools, and learning to filter relevant content, is crucial.

Promoting awareness of healthy technology use, educating about risks and benefits, and encouraging self-regulation are essential. Seeking professional support when necessary and practicing stress management techniques like meditation and mindfulness can improve mental health.

Limiting time on social media platforms and being selective about shared information can help mitigate social comparison and anxiety. Taking measures to protect online privacy, such as configuring privacy settings and being mindful of shared information, is vital.

Prioritizing face-to-face interactions and strengthening relationships outside the digital realm can counteract loneliness and improve emotional well-being. Addressing technological stress involves finding a healthy balance between technology and daily life, prioritizing mental and physical well-being in an increasingly digitized world.

The constant dependence on electronic devices can lead to anxiety related to the need to be always connected. The anticipation of messages, notifications, and the pressure to respond immediately can generate stress. Indeed, anxiety related to the need to be always connected is a common manifestation of technological stress in contemporary society. Some additional points that could further elaborate on this perspective include:

The fear of missing out on important things on social media or in digital communication can contribute to anxiety. Constant exposure to others' activities can create a sense of dissatisfaction and the feeling of not measuring up (FOMO).

The culture of immediate response and unrealistic expectations regarding availability can be overwhelming. The pressure to always be online can affect the quality of personal time and create conflicts with obligations outside the digital world.

Dependence on electronic devices before sleep can interfere with sleep quality. Exposure to blue light from screens can affect the sleep cycle, contributing to fatigue and stress.

Constant connectivity can blur the boundaries between work and personal life, making it difficult to disconnect. This can affect the balance between work and personal life, leading to stress and burnout.

The perceived need to maintain a positive online image can generate anxiety and stress. Constant comparison with others and seeking validation on digital platforms can negatively impact self-esteem.

Frequent notifications can disrupt daily activities and create a constant sense of urgency. This can affect concentration and productivity, contributing to work-related and personal stress.

Addressing these aspects involves not only setting technological boundaries, such as periods of disconnection but also promoting a culture that values mental well-being and encourages healthy technology use. Awareness and active management of the relationship with technology are key to mitigating the stress associated with constant connectivity.

The constant onslaught of information through emails, social media, and other digital platforms can lead to information overload. The need to process large amounts of data can contribute to stress and mental fatigue.

Information overload is a common consequence of constant exposure to a large amount of information through various digital platforms. Here are some additional aspects related to this challenge and how to address it:

Information overload can lead to decision fatigue. The constant need to evaluate and process information can deplete mental resources, making it harder to make informed and effective decisions.

With the vast amount of information available, it can be challenging to determine what is truly relevant and prioritized. This can contribute to procrastination and increase the feeling of overwhelm.

Frequent notifications and the uninterrupted flow of information can make it difficult to concentrate on important tasks. This can affect productivity and contribute to work-related stress.

The amount of information can make people feel like they don't have enough time to process everything. This can generate anxiety about constantly being "behind" or "missing out" on something important.

The abundance of information can increase the risk of exposure to misinformation. Additionally, the need to filter accurate information from inaccurate information can contribute to overthinking and anxiety.

Identifying and prioritizing the most relevant and urgent information can help manage information overload. Establishing specific times to check emails and social media can avoid constant interruption and provide periods of distraction-free focus.

Using filtering and organizational tools to manage information. Labeling emails, using reading lists, and categorizing information can facilitate access when needed.

Recognizing personal limits and learning to say no to non-essential information or tasks can help reduce mental burden.

Improving time management skills can help allocate time efficiently for processing information and completing tasks.

Incorporating mindfulness practices and mindfulness techniques can help reduce anxiety and improve the ability to focus on the present.

By adopting effective strategies to manage information overload, it is possible to reduce the associated stress and promote a healthier and more balanced use of technology.

Frequent interruptions due to notifications and messages can affect concentration and workflow. The need to constantly shift attention from one task to another can generate stress, especially in work environments.

Constant interruptions, especially those generated by notifications and digital messages, can have a significant impact on concentration, productivity, and overall well-being. Here are some additional aspects and strategies to address constant interruptions:

Continually shifting attention from one task to another can reduce efficiency and the quality of work. Interruptions can lead to the loss of contextual information and increase the time needed to refocus on the original task.

Frequent interruptions can generate stress and frustration, as individuals may feel pressured to respond quickly or feel that they cannot complete tasks effectively due to constant distractions.

"Flow" is a mental state of immersion in a task, and constant interruptions can make it challenging to enter and maintain this state. This can negatively affect creativity and the quality of work.

The need to repeatedly switch between tasks can deplete mental resources, contributing to mental fatigue and decreasing the ability for sustained attention.

Constant interruptions not only affect work performance but can also interfere with personal life. Difficulty in disconnecting can create tensions in relationships and affect time dedicated to personal and leisure activities.

Customize notification settings to prioritize important messages and reduce unnecessary distractions.

Define specific periods of the day as focus times, during which interruptions are minimized, and concentration on critical tasks is prioritized.

Clearly communicate moments when one is available to respond to messages and emails, setting realistic expectations for response times.

Use techniques like the Pomodoro Technique or time management apps to structure work into focused time blocks, followed by breaks.

Promote non-multitasking work practices, focusing on one task at a time to improve the quality and efficiency of work.

Establish work areas or specific times where interruptions are minimal, promoting an environment conducive to concentration.

In work environments, encourage a culture that respects and values time dedicated to important tasks, recognizing that frequent interruptions can negatively impact productivity.

By adopting these strategies, it is possible to mitigate the negative effects of constant interruptions and improve the quality of work and overall well-being.

Technology is often associated with performance pressure. Whether in the workplace, academia, or personal life, constant online visibility and social comparison can increase pressure to meet perceived standards, generating stress.

The association between technology and performance pressure is a reality in many aspects of life and can have a significant impact on mental and emotional health. Here are some additional aspects of this dynamic and proposed strategies for handling performance pressure associated with technology:

Social media and other online platforms often foster constant comparison with others. Carefully curated representations of life online can create additional pressure to meet perceived standards or match the apparent success of others.

Constant online visibility can create a sense of being under constant observation. This can increase pressure to maintain a positive image and meet performance expectations, as achievements and challenges can be shared publicly.

The culture of instant response in digital communications can generate anxiety and pressure to always be available. Delayed responses may be perceived as a lack of commitment or insufficient performance.

Technology often facilitates perfectionism, as it allows constant editing of images, messages, and achievements. This can increase pressure to reach perfectionist standards and contribute to emotional exhaustion.

The constant pressure to meet perceived standards can have negative consequences for mental health, including high levels of stress, anxiety, and the possible emergence of issues such as imposter syndrome.

Establish clear limits on the time spent on social media and other online platforms to reduce exposure to constant comparison.

Promote honesty and authenticity online, encouraging an understanding that representations on social media do not always reflect the complete reality.

Schedule regular periods of complete disconnection to reduce constant performance pressure and allow periods of mental rest.

Set achievable and realistic goals, recognizing that perfection is unrealistic, and each person has their own pace and trajectory.

Encourage practices that promote emotional well-being, such as mindfulness, meditation, and regular exercise, to counteract the negative effects of stress and performance pressure.

Seeking support from friends, family, or mental health professionals to share concerns and gain external perspectives is crucial.

Working on developing resilience skills to manage stress and challenges, recognizing that performance will not always be perfect, and errors are opportunities for learning.

Effectively managing performance pressure associated with technology involves adopting balanced and healthy practices that promote a realistic and sustainable approach to performance and personal authenticity.

The difficulty of disconnecting from electronic devices, especially outside working hours, can contribute to stress. The feeling of always being available can lead to exhaustion and negatively impact the balance between work and personal life.

The constant difficulty of disconnecting from electronic devices, especially outside working hours, is an increasingly common issue in modern society. This trend can have various impacts on mental health and overall well-being. Here are some additional aspects and specific strategies to address the difficulty of disconnection:

The feeling of always being available can create additional pressure, as immediate responses are expected, even outside working hours.

Difficulty disconnecting can affect personal relationships, as constant attention to electronic devices can reduce the quality of time spent with friends and family. The inability to disconnect can interfere with the ability to recharge and relax, contributing to physical and mental exhaustion.

The lack of disconnection can blur the boundaries between work and personal life, making it challenging to establish a healthy balance between the two.

Defining specific times to disconnect from electronic devices after work and during the weekend can help establish clear boundaries.

Designating a specific area at home as a device-free space can facilitate disconnection and encourage moments of rest.

Customizing notification settings to reduce constant interruption and allow moments of tranquility.

Communicating with colleagues, friends, and family about availability limits outside working hours can help manage expectations and reduce the pressure to always be connected.

Cultivating hobbies or activities that do not involve electronic devices can provide healthy alternatives for free time and facilitate disconnection.

Scheduling specific moments to spend quality time with friends and family without digital distractions can strengthen relationships and improve emotional well-being.

Creating specific bedtime routines, such as avoiding electronic screens, can facilitate better rest and improve sleep quality.

By adopting these strategies, it is possible to address the difficulty of disconnecting and promote a healthier balance between work and personal life, thus reducing stress associated with constant connectivity.

Prolonged exposure to digital screens can cause visual fatigue and postural problems. Physical stress from visual and postural discomfort can affect overall health and contribute to technological stress.

The difficulty of disconnecting from electronic devices, especially outside working hours, is an increasingly common problem in modern society. This trend can have various impacts on mental health and overall well-being. Here are some additional aspects and specific strategies to address the difficulty of disconnection:

The feeling of always being available can create additional pressure, as immediate responses are expected, even outside working hours.

Difficulty disconnecting can affect personal relationships, as constant attention to electronic devices can reduce the quality of time spent with friends and family.

The inability to disconnect can interfere with the ability to recharge and relax, contributing to physical and mental exhaustion.

The lack of disconnection can blur the boundaries between work and personal life, making it challenging to establish a healthy balance between the two.

Defining specific times to disconnect from electronic devices after work and during the weekend can help establish clear boundaries.

Designating a specific area at home as a device-free space can facilitate disconnection and encourage moments of rest.

Customizing notification settings to reduce constant interruption and allow moments of tranquility.

Incorporating mindfulness practices and techniques can help focus on the present and reduce anxiety associated with disconnection.

Communicating with colleagues, friends, and family about availability limits outside working hours can help manage expectations and reduce the pressure to always be connected.

Cultivating hobbies or activities that do not involve electronic devices can provide healthy alternatives for free time and facilitate disconnection.

Scheduling specific moments to spend quality time with friends and family without digital distractions can strengthen relationships and improve emotional well-being.

Creating specific bedtime routines, such as avoiding electronic screens, can facilitate better rest and improve sleep quality.

By adopting these strategies, it is possible to address the difficulty of disconnecting and promote a healthier balance between work and personal life, reducing stress associated with constant connectivity.

Visual and postural fatigue associated with prolonged use of electronic devices is a common problem in modern society, especially with the increasing dependence on technology. These issues can have a significant impact on physical health and contribute to technological stress. Here are some additional aspects and strategies to address visual and postural fatigue:

Prolonged exposure to digital screens can lead to computer vision syndrome, including symptoms such as blurred vision, dry eyes, headaches, and difficulty focusing.

Adopting uncomfortable postures during the use of electronic devices, such as hunching over the phone or sitting in a non-ergonomic position in front of a computer, can contribute to postural problems and muscle pain.

Visual and postural fatigue can affect productivity at work and in daily activities, as physical discomfort can distract and decrease concentration.

Prolonged exposure to screens before bedtime can affect sleep quality by interfering with the production of melatonin, a key sleep hormone.

Incorporating regular breaks during device use to rest the eyes and perform stretches to relieve muscle tension.

Adjusting the brightness, contrast, and font of screens to reduce visual fatigue and improve legibility. Using apps or settings that reduce blue light emission on screens during the late afternoon and evening to mitigate the impact on sleep.

Adjusting the height and position of the computer screen to ensure an ergonomic posture and using chairs and desks that promote good posture.

Incorporating simple eye exercises, such as frequent blinking and focusing on distant objects, to alleviate visual fatigue.

Considering the use of glasses designed to reduce visual fatigue and protect the eyes from blue light emitted by digital screens.

Maintaining proper eye hydration using artificial tears or blinking more frequently to relieve dry eyes.

Incorporating stretching exercises for the neck, shoulders, and back to prevent postural problems and relieve muscle tension.

Limiting exposure time to digital screens, especially before bedtime, to reduce the impact on sleep and visual fatigue.

Scheduling regular eye exams to ensure optimal visual health and address any visual problems early.

By implementing these strategies, visual and postural fatigue associated with the use of electronic devices can be reduced, improving physical health and contributing to mitigating technological stress.

The culture of instant responses associated with technology can create unrealistic expectations. The pressure to respond immediately to messages and emails can generate anxiety and stress due to the constant need to be available.

The pressure to provide immediate responses in the context of technology is a reality in modern society and can significantly contribute to technological stress. Here are additional aspects of this dynamic and strategies to address immediate response expectations:

The constant availability of electronic devices and internet connection can create the expectation of being available at any time of the day, contributing to a sense of lack of boundaries in personal and professional life.

The culture of immediate response can generate anxiety, as delayed responses may be perceived as a lack of commitment or a violation of social expectations.

The constant pressure to respond quickly can lead to emotional exhaustion, negatively impacting mental health and contributing to burnout.

The need to respond immediately can make it difficult to disconnect even during personal time, creating a constant sense of intrusion.

Clear and proactive communication about availability limits and setting realistic expectations for response times.

Establishing specific times to review and respond to messages, allowing periods of uninterrupted focus.

Configuring automatic responses in emails or messages to inform others when unavailable and when they can expect a response.

Promoting a culture that respects others' time and recognizes availability limits, both in the workplace and personally.

Learning to prioritize tasks and messages can help manage workload and respond effectively without feeling overwhelmed.

Scheduling specific times of the day to disconnect and take breaks can help reduce anxiety associated with the constant need to respond.

Encouraging mindful practices when using technology, recognizing that an immediate response is not always necessary and that it is beneficial to disconnect at certain times.

Maintaining strong personal boundaries and resisting external pressure to respond immediately, recognizing the importance of caring for mental and emotional health.

By adopting these strategies, it is possible to more effectively manage expectations of immediate response, thus reducing stress associated with the constant pressure to be always available.

Social media platforms often encourage constant comparison. Seeing the achievements and activities of others online can create stress, feeling the need to keep up or surpass perceived social expectations.

Online social comparison is a prevalent reality in the era of social media and can have a significant impact on people's emotional well-being. Here are additional aspects and strategies to address stress associated with online social comparison:

Constant comparison on social media can contribute to the fear of missing out on exciting experiences or important events that others are living.

Constantly viewing the seemingly successful lives of others online can lead to negative self-evaluation, lowering self-esteem, and generating feelings of dissatisfaction.

Exposure to perceived standards on social media can create pressure to conform to certain ideals, whether related to physical appearance, professional success, or personal life.

Constant comparison can contribute to anxiety, depression, and other mental health issues, especially when there is a perceived gap between one's own life and the seemingly perfect lives of others.

Developing awareness of one's own achievements, values, and personal goals, recognizing that each person's life is unique and not always accurately reflected on social media.

Setting time limits for social media use can help reduce exposure to constant comparison and promote a more balanced focus on real-life experiences.

Promoting a culture of authenticity on social media, where people feel comfortable sharing real experiences, including challenges and failures, rather than just highlighting positive aspects.

Strengthening connections in the real world can provide a more solid foundation for self-esteem and reduce dependence on online validation.

Cultivating gratitude by focusing on personal blessings and achievements can counteract feelings of dissatisfaction generated by constant comparison.

Taking occasional breaks from social media can provide perspective and reduce the pressure to keep up with others' lives.

Sharing feelings of discomfort or stress related to online social comparison with close friends, family, or mental health professionals can provide support and perspective.

Directing attention toward personal growth and individual progress instead of constantly comparing oneself to others can increase self-acceptance and well-being.

By adopting these strategies, stress associated with online social comparison can be mitigated, promoting a healthier and more balanced approach to social media use.

Lack of dedicated time to disconnect and relax away from electronic devices can contribute to technological stress. The inability to set clear boundaries can affect the quality of downtime.

Lack of time to disconnect can result in constant digital exhaustion, as individuals may feel always connected and available.

Lack of clear boundaries can lead to frequent interruptions during personal time, affecting the quality of time spent on leisure and rest activities.

Inability to disconnect can affect personal relationships by reducing the quality of time spent with friends and family.

An overloaded schedule can leave little room for downtime and disconnection, contributing to stress and a sense of overwhelm.

Recognizing the importance of downtime and making it a priority in the schedule by allocating specific times to disconnect.

Defining clear boundaries between work and personal time, avoiding the temptation to check emails or perform work tasks outside working hours.

Developing disconnection routines before bedtime, such as turning off electronic devices an hour before sleep, to facilitate better rest.

Scheduling specific moments to spend quality time with friends and family without digital distractions to strengthen interpersonal connections.

Scheduling specific days to completely disconnect from electronic devices, allowing uninterrupted periods of rest.

Taking advantage of vacations or days off to completely disconnect and dedicate time to relaxing and rejuvenating activities.

Incorporating well-being routines, such as regular exercise, meditation, or reading, to create regular spaces for disconnection and relaxation.

Evaluating the load of commitments and learning to say no to non-essential activities to free up time and reduce the feeling of overwhelm.

Being mindful of the time spent on electronic devices and setting conscious limits to ensure regular periods of disconnection.

In workplace environments, fostering a culture that values and promotes disconnection outside working hours to improve employees' quality of life.

By adopting these strategies, it is possible to address the lack of time to disconnect, thus promoting a healthier balance between technology and personal well-being.

Developing effective coping strategies is essential for managing technological stress. This may include setting time limits on device use, practicing digital disconnection at specific times, and seeking activities that promote well-being away from the screen.

Setting clear limits on the time spent on electronic devices. This may include establishing a daily limit for social media or general screen time.

Scheduling specific times of the day or week to completely disconnect from electronic devices. These disconnection periods allow for relaxation and rejuvenation away from digital distractions.

Establishing routines before bedtime that do not involve electronic devices. Turning off screens an hour before sleep can improve sleep quality and reduce visual fatigue.

Seeking outdoor activities that do not involve the use of electronic devices, such as walking, hiking, or engaging in sports. These activities promote well-being and provide a necessary break from technology.

Cultivating hobbies that do not require the use of electronic devices, such as reading, gardening, painting, or cooking. These hobbies can be therapeutic and offer a way to disconnect.

Scheduling specific days to completely disconnect from technology. Use these days to engage in activities that promote well-being and relaxation.

Incorporating mindfulness practices and meditation to reduce stress and improve concentration. Mindfulness can be practiced both in formal sessions and during everyday moments.

Prioritizing time spent on personal relationships outside the digital environment. Organizing face-to-face meetings and strengthening interpersonal connections can be beneficial for mental health.

Integrating regular physical exercise into the week. Exercise not only benefits physical health but can also be an effective way to reduce stress and improve mood.

Customizing notification settings on devices to minimize unnecessary interruptions. This allows for greater control over attention and reduces the pressure to respond immediately.

In workplace settings, participating in programs that promote disconnection outside working hours. This may include policies for time off and respect for availability limits.

If technological stress significantly impacts mental well-being, seeking professional support, such as therapy or counseling, can be beneficial for developing personalized coping strategies.

By adopting these strategies, a balanced approach to technology can be created, mitigating associated stress and promoting comprehensive well-being. It is important to adjust these strategies based on individual needs and make continuous adjustments as necessary.

Managing technological stress involves finding a healthy balance between utilizing technology and preserving emotional and physical well-being.

16.Technology and Loneliness

The relationship between technology and loneliness is an intriguing paradox that has emerged in the digital era. Although technology has facilitated instant connection and long-distance communication, it has also led to phenomena of loneliness. Despite being more digitally connected, some online interactions can be superficial. The lack of face-to-face contact can limit the depth of connections, contributing to a sense of emotional loneliness.

Superficial online connection is a common phenomenon in the digital era, where, despite being more connected, interactions often lack depth and authenticity. Here are some aspects of this phenomenon explored:

Online communication, often through text messages, emails, or social media comments, tends to be asynchronous. This lack of synchronization can make it challenging to grasp emotional nuances and build deeper connections.

Online interactions lack the richness of non-verbal communication present in face-to-face interactions. The absence of facial expressions, tone of voice, and body language can limit emotional understanding.

In digital environments, attention often focuses on self-presentation and the expression of achievements and positive aspects. This can lead to superficial interactions centered on external image rather than sharing deeper experiences.

The speed and ease of online connections can lead to a preference for quantity over quality. People may accumulate numerous superficial connections instead of cultivating a few deeper relationships.

Superficial connections may lack the capacity to provide meaningful emotional support. The absence of deep interactions can contribute to a sense of emotional loneliness, especially in challenging times. The lack of face-to-face contact can make expressing emotions and vulnerabilities uncomfortable, leading people to be more reluctant to share authentic personal experiences.

The superficial nature of many online interactions can contribute to negative social comparison. By observing only positive representations of others, people may feel that their own lives do not meet certain standards, thereby increasing the sense of loneliness.

Recommendations to address this issue include prioritizing face-to-face interactions whenever possible, fostering openness and vulnerability in online interactions, cultivating deeper relationships instead of pursuing a large number of superficial connections, actively participating in meaningful and personal conversations, and taking breaks from digital connectivity for moments of reflection and authenticity outside the online environment.

Social media often portrays an edited and highlighted version of people's lives. Constant comparison with these idealized representations can generate feelings of inferiority and isolation, contributing to loneliness.

Social comparison in the context of social media is a significant phenomenon that can have a profound impact on emotional health and contribute to loneliness.

People tend to share highlighted moments and positive aspects of their lives on social media, creating idealized representations. This selectivity can give the perception that others' lives are more successful or happier.

Constant comparison with the seemingly exciting lives of others can fuel the fear of missing out (FOMO). This fear can contribute to feelings of isolation and the worry of not measuring up.

Constant exposure to idealized images and narratives can affect self-image. Negative comparison can lead to feelings of inferiority and negative self-evaluation, increasing the sense of isolation.

Exposure to standards of beauty, success, and happiness on social media can create pressure to conform to certain ideals. Feeling inadequate can contribute to emotional loneliness.

Developing awareness that social media representations are not the complete reality, remembering that people selectively share, and acknowledging that life has both positive and challenging aspects is crucial. Setting time limits for social media use can reduce constant exposure to idealized images and mitigate social comparison.

Focusing on personal growth and individual achievements, celebrating one's successes, no matter how small, and cultivating authentic relationships outside of social media can counter feelings of dissatisfaction generated by constant comparison. Taking regular breaks from social media can help gain perspective and reduce the pressure to keep up with others' lives.

Developing a positive self-narrative by recognizing strengths and individual achievements, and avoiding excessive self-criticism, facilitates a more positive mindset. Encouraging a culture of authenticity on social media, where people feel comfortable sharing both positive aspects and challenges of their lives, can contribute to a healthier relationship with technology and online interactions.

Excessive reliance on digital communication can lead to the substitution of in-person interactions. The lack of direct social experiences can contribute to isolation and loneliness.

The substitution of in-person interactions with digital communication is a phenomenon that has emerged with the growing dependence on technology. Here are some aspects of this phenomenon explored, along with strategies to address it:

Digital communication, often through text messages, emails, or social media, tends to be asynchronous. This lack of synchronization can lead to emotional disconnection and the substitution of face-to-face interactions.

Constant immersion in electronic devices, especially in social settings, can result in technological isolation. People may be physically present but emotionally disconnected as they focus on their devices.

Excessive dependence on digital communication can lead to a reduction in direct social experiences. The lack of in-person interactions can contribute to social isolation and a sense of loneliness.

The substitution of in-person interactions with digital communication can limit the richness of human connection. The absence of facial expressions, eye contact, and other non-verbal cues can affect the quality of communication.

Consciously prioritizing face-to-face interactions, including social gatherings, group activities, and events that encourage direct connection, can be beneficial. Establishing clear boundaries for electronic device use in social and personal settings and designating specific periods of disconnection to facilitate in-person interaction are essential. Joining local groups or communities with similar interests promotes in-person interactions and the development of meaningful relationships.

Creating awareness of the impact of technology on social interactions and recognizing when it's appropriate to disconnect to fully engage in social life is important. Implementing disconnection routines before bedtime, turning off electronic devices an hour before sleep, can improve downtime quality and promote restful sleep. Proactively planning social encounters and activities with friends and loved ones can strengthen connections and reduce dependence on digital communication.

Working on the development of in-person communication skills, including active listening, eye contact, and the expression of emotions, is crucial for building meaningful relationships. By adopting these strategies, the trend of substituting in-person interactions with digital communication can be counteracted, fostering a more genuine and satisfying connection with others.

Constant immersion in electronic devices, especially in social settings, can lead to technological isolation. The lack of face-to-face interaction can contribute to loneliness, even in social situations.

Technological isolation is a phenomenon that arises when people constantly immerse themselves in electronic devices, emotionally disconnecting from their immediate social environment. Here are some aspects of this phenomenon explored, along with strategies to address it:

Excessive focus on electronic devices can lead to emotional disconnection from physically present individuals. Lack of attention and absence of face-to-face interaction contribute to technological isolation.

Individuals engrossed in their electronic devices may decrease their engagement in conversations and social activities. This creates barriers to interpersonal connection and can generate a sense of isolation even in social environments.

Technological isolation often manifests as a lack of awareness of the surrounding environment. People may be physically present, but their attention is directed at screens, limiting interaction with those around them.

Excessive dependence on technology can impact the quality of interpersonal relationships. Lack of face-to-face interaction can result in misunderstandings, reduced empathy, and weakened emotional bonds.

Establishing clear rules for electronic device use in social situations, setting time limits, and encouraging active participation in conversations and activities are important steps. Adopting a moderate approach to electronic device use, allocating specific times for checking messages and notifications, allows for greater presence in social environments.

Designating technology-free spaces or moments, where interactions are exclusively face-to-face, promotes interpersonal connection without digital distractions. Being mindful of the surrounding environment and the people present, practicing mindfulness helps regain awareness of the present moment and enhances connection with the social environment.

Prioritizing meaningful conversations and actively engaging in them contributes to building strong relationships and reduces technological isolation by focusing on the quality of interaction. Implementing disconnection routines before bedtime, turning off electronic devices an hour before sleep, not only improves sleep quality but also encourages interpersonal connection before bedtime.

Encouraging a culture that values face-to-face conversation and communication can be especially relevant in family, educational, and workplace settings.

Engaging in group activities or social events where face-to-face interaction is essential fosters connection and reduces technological isolation.

Providing education on healthy technology use, emphasizing the importance of balancing digital connectivity with active participation in social life.

Cultivating digital empathy by remembering that online interactions do not replace the authenticity and emotional connection of face-to-face relationships.

By adopting these strategies, it is possible to counteract technological isolation and promote increased interpersonal connection in social environments.

Digital communication sometimes lacks the authenticity and emotional connection experienced in face-to-face interactions. This can contribute to the feeling of loneliness, even when digitally connected.

The lack of authenticity in digital communication is a common challenge that can contribute to a sense of loneliness, even when digitally connected. Here are some aspects of this phenomenon explored, along with strategies to foster authenticity in online interactions:

The tendency to filter or edit content before sharing it on digital platforms can lead to selective representations of life. Lack of transparency can contribute to the perception that online interactions are superficial.

Fear of judgment or criticism online can lead people to present carefully crafted versions of themselves. Concern for social acceptance can reduce the genuine expression of thoughts and emotions.

Digital communication lacks many non-verbal cues present in face-to-face interactions, such as facial expressions and tone of voice. The absence of these cues can make conversations more prone to misunderstandings and less authentic.

Constant exposure to the seemingly perfect lives of others on online platforms can create the perception that one must maintain a similar image. This can lead to a lack of authenticity to meet certain perceived standards.

Prioritizing honesty in online communication. Being authentic about experiences, thoughts, and emotions can foster more genuine connections.

Recognizing and accepting vulnerability as a natural part of the human experience. Sharing personal experiences and challenges can build emotional bridges and reduce the feeling of loneliness.

Creating online communities that foster a supportive and accepting environment. This can encourage people to be more authentic by feeling secure to share their true experiences.

Remembering that online representations do not capture the complete reality. Avoiding constant comparison with others and focusing on personal growth can promote authenticity.

Engaging in platforms and online spaces that value and promote authenticity. Seeking communities that encourage people to be themselves without fear of judgment.

Practicing active listening in online interactions. Demonstrating empathy and understanding contributes to building more authentic connections.

Expressing emotions constructively and openly. Sharing joys and challenges can strengthen emotional connections and reduce the feeling of loneliness.

Establishing online spaces where people feel free to be authentic. This may include private discussion forums or closed groups where members can share more openly.

Promoting education on the importance of digital authenticity and the acceptance of diversity of experiences. This can contribute to a cultural shift towards honesty online.

Remembering that each person is unique, with individual experiences and perspectives. Celebrating the diversity and authenticity of each individual rather than conforming to predefined standards.

By incorporating these strategies, a more authentic and supportive digital environment can be promoted, thus reducing the feeling of loneliness associated with a lack of authenticity in online interactions.

Recommendations for addressing the "filter bubble" effect:

Actively seek out diverse and reliable sources of information. Explore newspapers, magazines, and websites that present different perspectives.

On some platforms, it's possible to disable content personalization in account settings. This can allow for a greater variety of information and opinions in the news feed.

Challenge yourself to explore content that doesn't align directly with your current opinions. Seek information that challenges and enriches understanding, even if it initially contradicts your beliefs.

Engage in discussions and online communities that address a variety of perspectives. Participating in constructive debates can broaden understanding and reduce the feeling of intellectual loneliness.

Be conscious about whom you follow on social media and other platforms. Actively seek out content that offers different and enriching perspectives.

Advocate for transparency in recommendation and personalization algorithms. Disclosing how these algorithms work can allow users to make more informed decisions about their content consumption.

By implementing these strategies, it is possible to counteract the filter bubble effect and encourage a greater diversity of perspectives in the online experience, thus reducing the feeling of intellectual loneliness.

Constant exposure to online information overload can be overwhelming. Occasional disconnection to avoid this overload may lead to a temporary perception of loneliness but is necessary for well-being.

Online information overload is a reality in the digital age, where constant exposure to a large amount of data can be overwhelming. Here are some aspects of this phenomenon explored, along with strategies to handle information overload:

Information flows continuously through various online platforms, such as social media, emails, and news. This constant flow can be challenging to process and may contribute to the feeling of being overwhelmed.

The abundance of available information can make it difficult to prioritize and select what is relevant. The lack of effective filters can lead to the feeling of being lost in a sea of data.

Information overload can have a negative impact on mental health, contributing to stress, anxiety, and mental fatigue. Constant exposure to negative news can also affect emotional well-being.

Define time limits for online information consumption. Establish specific times of the day to check news and social media to avoid excessive exposure.

Identify and prioritize reliable and relevant sources of information. Limit exposure to sensationalist or unverified sources to reduce information overload and promote the quality of consumed information.

Schedule regular periods of complete disconnection from electronic devices. This can help recharge the mind, reduce stress, and provide a necessary break from information overload.

Use filtering and organization tools to prioritize relevant content. Setting up alerts and selective notifications can help stay informed without feeling overwhelmed by the amount of information.

Develop discipline when consuming information. Be aware of personal limits and make informed decisions about what and when to access information to maintain a healthy balance.

Instead of seeking a large amount of information, focus on the quality of information. Select sources and content that are relevant and provide valuable information. Seek information from a variety of sources to gain diverse perspectives. Diversifying sources can help avoid the limitation of viewpoints that often contributes to information overload.

Set clear goals for information consumption. Establish daily or weekly limits to avoid saturation and enable a more intentional focus on seeking information.

Incorporate mindfulness practices to be present in the current moment and reduce anxiety related to information. Mindfulness can help manage emotional overload.

Enhance media literacy to critically evaluate the quality and relevance of information. Developing skills to discern between reliable and less reliable sources contributes to more conscious information consumption.

By adopting these strategies, it is possible to manage information overload more effectively and find a healthy balance between staying informed and preserving emotional well-being.

Prioritize building meaningful relationships online and offline. Engaging in online communities that share similar interests can be a way to foster genuine connections.

Establishing meaningful connections is essential to counteract the feeling of loneliness and promote emotional well-being. Here are some aspects of this strategy explored, along with recommendations for building meaningful relationships:

Human connections play a crucial role in emotional well-being. The quality of relationships is linked to personal satisfaction and the reduction of the feeling of loneliness.

Online communities offer opportunities to connect with people who share similar interests. Actively participating in these communities can facilitate the formation of meaningful relationships.

Striking a balance between online and offline connections is key. Both environments can contribute to the building of meaningful relationships, each with its own dynamics.

Seek and join online communities that focus on shared interests and passions. Identifying common interests facilitates connection and the building of meaningful relationships.

Actively participate in conversations and activities within online communities. Sharing experiences, ideas, and knowledge contributes to the formation of genuine connections.

To be authentic and open in sharing experiences and emotions fosters emotional connection and facilitates the building of meaningful relationships. Not neglecting offline relationships is crucial for emotional well-being. Maintaining and strengthening connections with friends, family, and colleagues significantly contributes to emotional well-being. Seeking conversations beyond the superficial and addressing meaningful topics fosters the building of substantial relationships.

Offering support and empathy to others in online communities strengthens bonds and creates stronger relationships. Organizing or participating in both online and offline events provides opportunities for more personal connections. Encouraging respect for diverse opinions and perspectives enriches relationships and promotes mutual understanding. Setting realistic goals for relationship building, such as making new connections each month or regularly participating in social activities, is important.

Valuing and allocating time for in-person interactions is essential, as face-to-face relationships often have a deeper and lasting impact on building meaningful connections. By following these recommendations, one can build and strengthen meaningful connections both online and offline, contributing to a sense of connection, belonging, and emotional well-being.

Being aware of the selective nature of representations on social media and limiting comparisons with others is essential. Focusing on personal growth rather than constant comparison can reduce loneliness. Limiting social comparison is fundamental to preserving emotional well-being and reducing the feeling of loneliness.

Social media often presents an idealized and selective version of people's lives. Comparing oneself to these representations can contribute to feelings of dissatisfaction and loneliness. Shifting focus toward personal growth and individual development, rather than measuring success against others, can be more rewarding. Recognizing that what is shown online does not represent the entirety of a person's life is crucial.

Setting time and frequency limits for social media use can reduce the feeling of loneliness and improve self-esteem. Celebrating personal achievements instead of constantly comparing oneself to others contributes to greater personal satisfaction. Adopting a growth mindset that values learning and progress over static comparison is beneficial. Avoiding destructive competition with others and instead collaborating and learning from their experiences and successes is important.

Scheduling regular periods of digital disconnection allows time to reconnect with oneself, reflect on personal achievements, and set realistic goals. Cultivating an attitude of gratitude by focusing on the positive aspects of one's life can reduce the feeling of loneliness. Setting sustainable and values-aligned personal goals can provide a sense of purpose and self-connection. Prioritizing authentic and meaningful relationships over superficial comparisons contributes to emotional satisfaction. Seeking support from friends, family, or other loved ones and sharing experiences and challenges can generate understanding and solidarity, reducing the feeling of loneliness.

By adopting these recommendations, one can shift focus toward a more positive and growth-centered mindset, thereby reducing social comparison and the feeling of loneliness.

Promoting face-to-face communication whenever possible, through video conferences, phone calls, or in-person meetings, helps build more authentic connections. Fostering face-to-face communication is essential for building authentic connections and reducing the feeling of loneliness.

Face-to-face communication allows for more authentic expression of emotions, as non-verbal signals contribute to a more comprehensive understanding. Face-to-face interactions tend to facilitate deeper and more meaningful connections, as visual contact, facial expressions, and other communicative nuances contribute to a richer understanding.

Where possible, choosing video conferences over text messages or emails enhances the quality of communication. Regularly scheduling phone calls with friends, family, or colleagues enriches communication and helps maintain stronger connections. When circumstances allow, organizing in-person meetings is crucial for building and maintaining authentic relationships.

Attending social events, gatherings, or activities in person provides opportunities for face-to-face interaction and building connections beyond online interactions. Prioritizing quality in-person time with friends and loved ones strengthens emotional bonds. In face-to-face conversations, encouraging openness and sincerity builds trust and authenticity. Occasionally disconnecting from electronic devices allows more time for face-to-face connections, and being fully present in the moment strengthens relationships. Participating in networking events or social activities where new face-to-face connections can be established creates opportunities for meaningful relationships. In face-to-face interactions, practicing active listening by paying mindful attention to the other person and demonstrating interest contributes to more effective communication. Establishing safe and trusting spaces for sharing thoughts and feelings in person strengthens emotional intimacy.

Promoting face-to-face communication, whether through digital means or in-person encounters, is crucial for building more authentic connections and reducing the feeling of loneliness. Incorporating these recommendations into daily interactions contributes to strengthening emotional and social bonds.

Establishing regular periods of disconnection to balance digital presence with time dedicated to offline activities and in-person relationships is crucial. Disconnecting to reconnect is a valuable strategy for balancing digital presence and fostering more meaningful connections with offline activities and in-person relationships.

Temporarily disconnecting from the digital world allows for restoring balance between online life and offline experiences, promoting overall well-being. Establishing regular disconnection periods helps reduce technological stress by providing a break from constant digital notifications and demands. Scheduling specific times of the day or week without the use of electronic devices, including mornings or evenings, can contribute to a healthier balance.

Designating technology-free zones at home, such as relaxation areas or family gathering places, creates a conducive environment for in-person interaction. Regularly designating full days of digital disconnection allows time exclusively for offline activities and in-person relationships.

Engaging in activities that do not require electronic devices, such as outdoor walks, reading physical books, or participating in sports, promotes disconnection and well-being. Making agreements with friends, family, or colleagues to disconnect together during certain moments strengthens relationships and creates shared experiences outside the digital world. Integrating daily or weekly routines that do not involve electronic devices provides specific moments for disconnection and focuses on more analog activities. Practicing mindfulness or mindfulness during disconnection moments contributes to a more meaningful experience in offline activities. Setting limits on notifications to reduce constant interruptions allows for more control over screen time and promotes disconnection.

Promoting education on healthy technology use in family and workplace environments raises awareness about the importance of disconnection to maintain balance. Regularly evaluating the effectiveness of disconnection periods and making adjustments as needed allows for flexibility in implementation to adapt to changing needs.

By incorporating these recommendations, one can cultivate a more balanced approach to technology and strengthen connections outside the digital realm, contributing to greater emotional and social well-being.

Actively engaging in social activities, clubs, or events in the community to foster face-to-face encounters and build meaningful relationships outside the digital realm is essential. Participating in social activities is an excellent way to encourage face-to-face encounters and build meaningful relationships outside the digital realm. Here are some aspects of this strategy and recommendations for participating in social activities that provide opportunities for in-person interactions, contributing to the construction of richer and more authentic relationships.

Involvement in social activities allows for enjoying a variety of experiences and meeting people with different interests, fostering diversity in social connections. Identifying personal interests and seeking related social activities facilitates connections with like-minded individuals.

Seeking local clubs or groups focusing on specific hobbies provides the opportunity to meet people who share similar passions. Participating in community events, fairs, or local festivals creates a conducive environment for social encounters and relationship building within the community.

Exploring volunteering opportunities in local organizations not only contributes to community well-being but also facilitates interaction with people committed to similar causes. Joining sports groups or exercise classes promotes physical health and offers opportunities to connect with individuals interested in well-being. Exploring participation in religious or spiritual groups, if of personal interest, provides opportunities for social connection based on shared values. Using online platforms like Meetup to find events and local meetings related to specific interests allows for informal meetings with new people.

Organizing social gatherings with friends, colleagues, or neighbors provides the opportunity to strengthen existing relationships and create new connections. Attending cultural events such as exhibitions, concerts, or theatrical performances enriches culturally and offers occasions for social interactions. Establishing regular social activities in the routine, such as game nights, dinner outings, or group walks, strengthens social connections through consistent participation.

By actively participating in social activities, an environment is created that encourages meaningful connections, builds authentic relationships, and enjoys the richness of face-to-face interactions outside the digital realm.

Promoting education on healthy technology use and awareness of the effects of online social comparison to mitigate loneliness is crucial. Promoting digital education is key to encouraging healthy technology use and increasing awareness of the effects of online social comparison, contributing to mitigating the feeling of loneliness. Here are some aspects of this strategy and recommendations for promoting digital education:

Digital education aims to raise awareness of how technology use affects mental health and social relationships, emphasizing the importance of a balanced approach. Digital education includes strategies to prevent online social comparison by providing tools to critically interpret information presented on digital platforms. Integrating digital education programs into school curricula teaches students about responsible technology use and managing online social pressure.

Offering awareness sessions on healthy technology use in workplace environments addresses the importance of balancing online work and disconnection. Organizing workshops for parents on how to guide their children in proper technology use and addressing challenges related to social comparison. Implementing awareness campaigns in local communities about the impacts of technology on mental health and social relationships.

This may include talks, community events, and educational resources. Providing online resources, such as blogs, articles, and educational videos, addressing topics related to mental health in the digital environment and managing social comparison.

Educating about conscious use of social media, focusing on authenticity and understanding that online representations do not reflect the entirety of a person's reality. Incorporating the teaching of digital literacy skills that include the ability to critically evaluate online information and develop strategies to manage technological stress. Offering counseling or digital therapy sessions to specifically address emotional challenges related to technology use and social comparison online.

Collaborating with mental health organizations and experts to develop and promote educational programs on digital well-being. Keeping digital education updated to reflect changes in technology and new challenges that arise. Regular updating ensures the relevance and effectiveness of educational programs. By promoting digital education, individuals are empowered to make informed decisions about their use of technology, reduce online social comparison, and mitigate the feeling of loneliness by fostering a balanced and healthy approach to digital life.

Actively seeking diversity of perspectives and opinions online to avoid the formation of bubbles and enrich the digital experience is essential. Supporting the diversity of perspectives online is crucial to avoid the formation of bubbles and enrich the digital experience. Here are some aspects of this strategy and recommendations to support the diversity of perspectives:

Actively seeking sources of information and opinions that represent a variety of perspectives, avoiding homogeneity of online information. Supporting the diversity of perspectives enriches the digital experience by offering a more complete and nuanced view of discussed topics. Following and diversifying sources on social media. Incorporating different perspectives contributes to a broader understanding of issues. Joining online groups that encourage respectful discussion and present a variety of opinions. Participating in constructive debates enhances understanding of different perspectives.

Consuming news and content from diverse media with varied editorial approaches helps avoid polarization and provides a balanced view of events. Engaging in intercultural conversations online. Actively seeking interaction with people from different cultures and backgrounds enhances the understanding of the world. Contributing to the diversity of online content by creating and promoting materials that represent a variety of perspectives. This can include blogs, videos, or social media posts.

Questioning assumptions and inherent biases in online information. Developing a critical attitude toward information contributes to a more objective view. Participating in open debate platforms that encourage diversity of opinions and promote respectful discussion. This broadens the range of considered perspectives. Engaging in initiatives that promote intercultural education and global understanding.

This includes supporting educational projects that seek to bridge cultural gaps and promote inclusion. Connecting with online communities that represent diverse identities, experiences, and opinions. Active participation in these communities fosters appreciation for diversity. Supporting and consuming media that practices quality journalism and presents multiple perspectives in its reports. Contributing to the support of responsible and equitable media.

By adopting these recommendations, one can contribute to building a more inclusive digital environment, where the diversity of perspectives is valued and the formation of information bubbles is avoided. This not only enriches the online experience but also promotes understanding and mutual respect.

If technological loneliness becomes overwhelming, seeking support from mental health professionals can be beneficial in addressing underlying emotional concerns. Recognizing the presence of technological loneliness and the need to address underlying emotional concerns. Seeking support from mental health professionals trained to address the emotional complexities related to technological loneliness. Seeking the help of a therapist or counselor specializing in mental health. These professionals can provide guidance and support in addressing emotional concerns. Consulting with psychologists specialized in technology and mental health. Some professionals have specific expertise in understanding and addressing challenges associated with technology use. Participating in online therapy sessions if location or time constraints are barriers to in-person care. Online therapy offers flexibility for receiving support.

Exploring online support groups where experiences can be shared with individuals facing similar challenges. These groups provide a space for connection and mutual support. In workplace settings, considering providing access to mental health professionals as part of employee wellness programs. Conversing with a psychiatrist if experiencing more intense or persistent emotional symptoms.

Psychiatrists are trained to assess and prescribe treatment if necessary. Working with a mental health professional to assess underlying factors contributing to technological loneliness and developing personalized strategies to address them. Learning and developing effective coping skills under the guidance of a professional.

These skills can help manage technological stress and improve emotional well-being. Considering specific therapies, such as cognitive-behavioral therapy, which focuses on changing negative thought and behavior patterns associated with loneliness and technological stress. Maintaining open communication with the mental health professional about specific challenges related to technology and loneliness. This facilitates a personalized approach to treatment.

Seeking professional support is a valuable step in addressing emotional concerns related to technological loneliness. Mental health professionals are trained to provide guidance, tools, and strategies that help improve emotional well-being and develop a healthier approach to technology use.

17.Impact of Video Games on Mental Health

The impact of video games on mental health is a complex and multifaceted topic that has been the subject of research and debate. Some studies suggest that certain types of video games can enhance cognitive skills such as attention, memory, and decision-making. Action games, requiring sustained and selective attention, have been observed to improve players' ability to track objects in a busy visual field.

Some video games, especially those involving puzzles and strategy, may contribute to the development of working memory. Working memory is essential for retaining and manipulating information in the mind while performing tasks. Action and strategy games, often requiring quick and precise decisions, have been associated with improved decision-making ability.

Strategy and simulation games can enhance multitasking abilities as players need to manage multiple tasks simultaneously to succeed in the game. Action games involving fast and precise movements can improve hand-eye coordination and manual dexterity. Puzzle and adventure games often require problem-solving skills as players overcome obstacles to progress in the game.

Real-time strategy games can improve strategic planning and long-term decision-making. It's important to note that while there is evidence of cognitive benefits associated with certain video games, not all games have the same impact, and the amount of time dedicated to gaming is also a crucial factor. Additionally, the effects may vary depending on the individual and other contextual factors.

In addition to cognitive aspects, social, emotional, and physical aspects related to video game use should also be considered. In summary, balance and moderation are key to ensuring that the enjoyment of video games does not have a negative impact on other aspects of daily life.

Online games can provide a space for socialization and community building, especially important for people who may have difficulty connecting in other ways. Socialization through online games is a significant phenomenon and can have positive impacts on community building. Here are some aspects related to socialization in online games:

Online games often lead to the formation of virtual communities where players from around the world can connect, interact, and collaborate in shared virtual environments.

The real-time nature of many online games allows direct interaction between players. They can communicate through voice chat, instant messaging, or even in-game video conferencing.

Online games offer opportunities for collaboration and competition among players. Collaboration can foster teamwork, while friendly competition can strengthen connections between participants.

Specific platforms are designed for social gaming where players can connect, join team matches, and form online friendships.

Online games can be inclusive, providing a space where people from diverse geographical locations, cultural backgrounds, and abilities can come together without the limitations of physical space.

Online gaming communities often provide a space for emotional support. Players can share experiences, express emotions, and find understanding among peers.

Some online games organize events and community activities, such as tournaments, online parties, or collaborative challenges, strengthening social ties and creating a sense of belonging.

Participating in online games can contribute to the development of social skills, such as communication, empathy, and conflict resolution, as players interact in virtual social environments. For individuals who may have difficulty socializing in face-to-face situations, online games offer an accessible and comfortable alternative for building relationships and friendships.

Connections formed through online games can be long-lasting, with some friendships beginning in the virtual environment and extending into real life. It is essential to consider that, while socializing in online games can have many positive aspects, it is also necessary to balance the time spent on these games with other social activities and responsibilities in daily life. Additionally, online safety and moderation in interaction are crucial aspects to ensure positive and respectful experiences in these virtual environments.

Video game addiction, especially to online games, is a concern. The inability to control the time spent on games can negatively affect daily life. Video game addiction, also known as Gaming Disorder, is a recognized concern by mental health experts and organizations such as the World Health Organization (WHO). Here are some aspects related to video game addiction:

Video game addiction is characterized by a persistent and recurrent pattern of behavior related to digital gaming. It can manifest as a loss of control over gaming, prioritizing gaming over other daily activities, and continued gaming despite negative consequences.

Video game addiction can be related to various factors, including the search for escape, the instant gratification provided by games, online socialization, and competition.

Online games, especially those that encourage social interaction and competition, may have a higher potential for addiction due to their continuous and often uninterrupted nature.

Video game addiction can negatively impact individuals' daily lives, including academic or work performance, interpersonal relationships, physical and mental health, and time management.

Some signs of video game addiction include a lack of control over gaming time, prioritizing gaming over other responsibilities, loss of interest in previously enjoyed activities, and persistence in gaming despite negative consequences.

Video game addiction can be diagnosed by mental health professionals. Treatment may include psychological interventions, cognitive-behavioral therapy, family support, and, in more severe cases, participation in specialized rehabilitation programs.

Prevention of video game addiction includes education on healthy gaming, emphasizing the importance of balancing game time with other activities, and early identification of warning signs.

The gaming industry has a responsibility to provide tools for time control and addiction management. Parents also play a crucial role in monitoring and setting limits on their children's gaming time.

It is essential to address video game addiction with empathy and understanding, recognizing that it can affect people of all ages. Awareness, education, and access to support resources are crucial to mitigate the risks associated with video game addiction and promote healthy technology use.

While online games can facilitate socialization, they can also lead to social isolation if players prefer spending time in the virtual world instead of interacting face-to-face. Social isolation is an important aspect to consider when exploring the relationship between online games and social interaction. Here are some points related to this impact:

Some players may prefer to spend a significant amount of time in the virtual world of online games instead of engaging in face-to-face social activities. This preference can lead to social isolation if it becomes a constant choice.

Time spent on online games may result in a reduction of in-person social interactions. If players choose to spend the majority of their time online, they may limit opportunities for face-to-face interaction with friends, family, or others in their environment.

The preference for the virtual world can affect interpersonal relationships. Friends and family may feel excluded or neglected if an individual spends an excessive amount of time in online games instead of participating in shared activities.

In the case of children and teenagers, social isolation related to online games can influence social development. Face-to-face interaction is essential for the development of social and emotional skills.

While online games offer opportunities for socialization, these interactions may be limited in terms of diversity. Depending on the chosen games, players may primarily interact with people who share similar interests, limiting exposure to new perspectives and experiences.

In extreme cases, the preference for online games can lead to excessive isolation, where individuals choose to withdraw almost completely from social interactions outside the virtual world.

The key is to find a healthy balance between time spent on online games and in-person social interactions. Social isolation can be reduced by maintaining a variety of social and recreational activities in daily life. Players should be aware of their behavior patterns and be willing to self-regulate their online time to avoid social isolation. Self-reflection and open communication with friends and family are important.

In summary, while online games offer opportunities for socialization, it is crucial to balance these interactions with participation in face-to-face social activities. Awareness, self-regulation, and open communication are key elements to mitigate the risk of social isolation associated with excessive use of online games.

Intensive involvement in gaming, especially during the night, can affect sleep patterns and contribute to mental health problems such as fatigue and irritability. The impact on sleep is an important aspect to consider when examining the relationship between video games, especially those played intensively at night, and mental health. Here are some points related to this impact:

Exposure to the light from video game screens before bedtime can negatively affect sleep quality. The blue light emitted by screens can interfere with the production of melatonin, a key hormone for regulating sleep.

Playing intensively before bedtime can increase mental excitement and stimulation, making it harder to fall asleep. The mind may remain active due to the excitement or concentration on the game.

Gaming late into the night can disrupt the circadian rhythm, the body's natural cycle that regulates sleep and wakefulness. This can lead to a misalignment of the internal biological clock.

Intensive involvement in games during the night may result in a reduction in the quantity and quality of sleep. Lack of adequate sleep can contribute to mental health problems such as fatigue, irritability, and difficulty concentrating.

Chronic sleep deprivation may be associated with mental health issues, including anxiety and depression. Proper sleep is essential for emotional and cognitive well-being.

Establishing a regular sleep routine, limiting screen exposure before bedtime, and creating a sleep-conducive environment are recommendations to improve sleep quality.

Gamers and their families should be aware of the importance of sleep hygiene, which includes practices and habits that promote healthy and restful sleep.

Setting limits on the time spent on video games before bedtime can be beneficial to allow the mind to relax and prepare for sleep. Lack of adequate sleep can affect daily performance, both academically and professionally, contributing to mental health problems related to stress and pressure.

Intensive involvement in video games, especially during the night, can have repercussions on sleep and, consequently, mental health. Awareness of the importance of sleep, along with the implementation of healthy sleep practices, is essential to mitigate these negative impacts and promote overall well-being.

Exposure to violence in some games can desensitize players, which could have implications for their perception of real-life violence. Desensitization to violence is a phenomenon studied in relation to constant exposure to violent content, including that found in some video games. Here are some aspects related to this phenomenon:

Desensitization involves a decrease in sensitivity or emotional response to specific stimuli due to repeated exposure to those stimuli. In the context of video games, it refers to a reduction in the player's emotional reaction to violence.

Some video games contain graphic and detailed representations of violence, ranging from combat to war scenes. Constant exposure to these representations can affect the player's perception.

Desensitization to violence occurs when players repeatedly experience violent situations in games and, as a result, may become less sensitive or emotionally reactive to such situations.

Desensitization could have implications for the perception of real-life violence. Desensitized players may have a lower emotional reaction or empathy towards real violence, having repeatedly experienced violent situations in a virtual context.

The relationship between exposure to violent content in video games and desensitization, as well as its impact on the perception of real-life violence, is a topic of debate. Some studies suggest correlations, but establishing a precise causal relationship is complex.

The reaction to violence in video games and the possibility of desensitization can vary based on individual differences such as personality, age, and frequency of exposure.

The context in which violence is experienced in video games is also important. The representation of violence within a moral or narrative context can influence how it is perceived.

Awareness of the potential effects of desensitization and education about distinguishing between virtual violence and reality are important. Players should be aware of how video games can affect their perception and behavior.

In summary, while desensitization to violence has been observed in some studies, the exact relationship between exposure to violent content in video games and its impact on the perception of real-life violence is complex and subject to various variables. Ongoing research and critical awareness are essential to better understand these phenomena.

There is a debate about whether violent games can contribute to real-life aggressive behavior, although the evidence is mixed and inconclusive.

The impact of violent games on aggressive behavior has been a topic of debate and has generated considerable research. Here are some key aspects related to this debate:

The evidence on whether violent games contribute to real-life aggressive behavior is mixed and often contradictory. Some studies suggest an association, while others do not find a direct relationship.

Meta-analyses and reviews of scientific literature have yielded divergent results. Some suggest a modest relationship between exposure to violent games and aggression, while others find that the evidence is not strong enough to establish a clear connection.

It is recognized that individual and contextual differences play a significant role in the relationship between violent games and aggression. Factors such as personality, family environment, and gaming frequency can modulate observed effects.

Some studies have shown short-term effects, where immediate exposure to violent games can temporarily increase aggression. However, whether this translates into long-term aggressive behavior is less clear.

Methodological challenges, such as the difficulty of controlling all factors that could influence aggression, have been noted in research on this topic. Additionally, diversity in types of games and a lack of consensus in aggression measures complicate result interpretation.

Some studies have explored theories such as catharsis, suggesting that violent games can serve as a way to release accumulated aggression. Others focus on modeling, arguing that observing violent behaviors in games could influence imitation.

While some professional associations and research groups have expressed concerns about the relationship between violent games and aggression, there are also experts and organizations that have pointed out the lack of conclusive evidence and advocated for a more balanced perspective.

Some researchers argue that, compared to other factors such as violence at home or exposure to violence in the media, violent games might have a relatively minor impact on aggression.

Ultimately, the debate on the impact of violent games on aggressive behavior continues, and research in this field remains an active area of study. Understanding this issue is complex and subject to a variety of carefully considered factors.

In online gaming environments, social pressure to excel and compete can generate stress and anxiety, especially in competitive games.

Competition and comparison in online gaming environments can have a significant impact on players' experiences and contribute to stress and anxiety. Here are some aspects related to this topic:

In online gaming environments, especially those that are highly competitive, there may be social pressure to excel and demonstrate exceptional skills. This pressure can generate anxiety as players feel the need to meet certain standards or expectations.

In competitive online games, the competition can be intense, and the feeling of being constantly evaluated by other players can increase pressure and stress. Concerns about performance can contribute to higher levels of anxiety.

Online environments often encourage constant comparison, whether in terms of skills, achievements, or equipment. This comparative environment can lead to unhealthy competition and generate insecurities among players.

Performance expectations can be high, especially in games with rankings and league systems. The pressure to maintain or improve one's position in the player hierarchy can generate performance-related anxiety.

In highly competitive environments, hostility and toxicity among players can be common. Exposure to negative behaviors can increase stress and negatively impact the enjoyment of the game.

Constant comparison and pressure to excel can affect players' self-esteem. Those who feel they do not meet established standards may experience anxiety and feelings of inadequacy.

It is crucial to promote healthy gaming and establish clear limits in terms of time and emotional commitment. Encouraging a more relaxed and enjoyable approach to gaming can help reduce pressure and stress associated with it.

Developing effective coping strategies, such as stress management, setting realistic goals, and occasional disconnection, can help players deal with the pressure and anxiety associated with competition in online games.

In conclusion, although online games can offer exciting and social experiences, intense competition and constant comparison can generate stress and anxiety.

Promoting a balanced approach, healthy gaming, and emotional support can contribute to a more positive experience in these environments.

Players may experience a direct relationship between their in-game performance and their self-esteem, which can have implications for mental health if they perceive themselves as "failures" in the game.

The relationship between in-game performance and self-esteem is an important aspect that can affect players' mental health. Here are some aspects related to this dynamic:

For many players, their in-game performance may be closely linked to their identity and self-concept. Experiencing success in the game can strengthen self-esteem, while perceived failure can negatively affect it.

In online gaming environments, self-evaluation is constant. Players can measure their worth and abilities through statistics, rankings, and achievements. This can create a significant emotional connection between performance and self-esteem.

Performance pressure can be intense, especially in competitive games. Players may feel the need to constantly prove their worth through their skills and achievements in the game, leading to stress and anxiety.

In-game performance can influence players' perception of themselves. Self-image can be affected positively or negatively based on successes or failures in the game, subsequently affecting self-esteem.

Perceived failure in the game can have a significant impact on self-esteem. If a player sees themselves as a "failure" in the game, this can affect their broader self-concept and generate negative emotions. Constant competition and comparison with other players can intensify the connection between in-game performance and self-esteem. Feeling outperformed by others can create insecurities and reduce self-esteem.

It is crucial to encourage a healthy perspective on in-game performance. Self-esteem should not depend solely on success in the game, and it's important to remember that games are just one aspect of life.

Encouraging players to diversify their sources of self-esteem is essential. Recognizing achievements outside of the game, such as personal skills, relationships, and academic or professional accomplishments, can help balance self-evaluation.

The importance of social and emotional support is crucial. Having connections outside the gaming world can provide valuable emotional backing and remind players that their worth is not limited to their performance in the game.

In summary, the relationship between in-game performance and self-esteem is complex and can have implications for mental health. Promoting a balanced perspective, diversifying sources of self-esteem, and providing social and emotional support are important strategies for addressing this dynamic in a healthy way.

It is essential to establish time limits for gaming and balance virtual activities with social interactions and other responsibilities. Setting time limits for gaming is a crucial practice to ensure a healthy and balanced use of video games. Here are some aspects related to setting limits:

Setting time limits promotes healthy use of video games by avoiding overexposure and excessive time dedicated to this activity.

It contributes to preventing gaming addiction, especially in situations where the line between entertainment and compulsion can become blurry.

It allows players to balance the time spent on video games with other important activities, such as work, studies, social interactions, and physical exercise. Time limits ensure that players fulfill their daily responsibilities and prevent gaming from interfering with important obligations like work, school, or family responsibilities.

Extended gaming sessions can lead to mental and physical fatigue. Setting limits helps prevent fatigue, which, in turn, can reduce the stress associated with marathon gaming sessions.

By limiting the time spent on video games, participation in in-person social interactions is encouraged. This is crucial for maintaining meaningful connections outside the virtual world.

Integrating time limits into the daily routine helps establish healthy habits and promotes a balance between gaming time and other important activities.

In family or shared environments, it is important to communicate and agree on time limits. Setting clear expectations helps prevent conflicts and encourages responsible use.

Players should engage in continuous self-assessment to ensure that the established limits are suitable and adjust them as needed to maintain a healthy balance.

In family environments with children, implementing parental control tools can help monitor and limit gaming time effectively.

In summary, setting time limits for gaming is a key strategy to ensure a balanced and healthy use of video games. This not only contributes to individual well-being but also promotes a...

Promoting open conversations about the time spent on video games, game choices, and the importance of a variety of activities in daily life. Encouraging open conversations about the time spent on video games is an effective strategy to promote a healthy and balanced use of this form of entertainment. Here are some aspects related to these conversations:

Establishing an open and respectful communication environment is crucial. Players should feel comfortable sharing their experiences and concerns without fear of judgment. When discussing video games, it's important to understand the interests and motivations behind the game. This can help identify positive benefits and address potential challenges.

Recognizing and highlighting the positive aspects of video games, such as the development of cognitive skills, online socialization, and fun, can contribute to a more balanced conversation.

Promoting awareness of the importance of a variety of activities in daily life is essential. Video games should be considered as part of the routine, not the only significant activity.

Setting clear expectations about the allowed time for playing and the appropriate moments to do so can help avoid conflicts and ensure responsible use.

Encouraging the exploration of other interests and activities outside the gaming world can broaden experiences and promote balance in everyday life.

Discussing the content of video games is important, especially in the case of children and teenagers. Ensuring that games are age-appropriate and aligned with family values is crucial.

Maintaining an adaptable and flexible attitude is key. Conversations should be an ongoing dialogue that adapts as interests and circumstances change.

Actively participating in gaming experiences, whether by playing together or showing genuine interest, can strengthen the connection and facilitate more open conversations.

Educating about healthy video game use and signs of potential issues, such as addiction, contributes to a more comprehensive understanding and promotes informed decisions.

Recognizing the importance of rest and leisure time is essential. Video games can be part of a healthy leisure time, as long as they are balanced with other activities.

In summary, open conversations about the time spent on video games are essential to create a healthy gaming environment.

The response to video games can vary depending on personality, age, and other individual factors. Additionally, the context in which video games are played can also influence their impact. Individual and contextual factors play a significant role in the experience of video games. Here are some aspects related to these factors:

The response to video games can vary considerably based on each individual's personality. Some people may find them exciting and rewarding, while others may experience negative emotions. Personalizing the experience is related to the diversity of individual preferences and responses.

Age also influences how video games are perceived and experienced. Children, teenagers, and adults may have different needs and reactions. It is important to consider the appropriateness of games based on the developmental stage.

There are disparities in game participation and preferences between genders. Some studies suggest differences in how men and women play and experience video games. Gender diversity in the gaming industry also influences the creation of diverse content.

Gaming style and individual preferences are key factors. Some people enjoy deep narratives, while others prefer competitive challenges. The diversity of genres and game types allows for tailoring the experience to each player's preferences.

The family and social environment in which video games are played is also relevant. Family involvement, online and offline social interactions, and the social perception of video games can influence the overall experience.

The mood and mental health of an individual can affect the gaming experience. Some people turn to video games as a way to escape or manage stress, while others may experience negative effects on their mental well-being.

Expectations and motivations for playing are crucial. Some play for fun and entertainment, while others seek achievements and challenges. Expectations can affect how achievements and failures in the game are perceived. Conditions in which video games are played, such as the physical environment and equipment used, also influence the experience. Factors like lighting, comfort, and the quality of the connection can affect immersion and enjoyment.

Cultural differences and social norms also play a role in the gaming experience. What may be considered acceptable or exciting in one culture may vary in another.

Considering these individual and contextual factors is essential to understand the diversity of experiences with video games and ensure an equitable and respectful approach to this form of entertainment.

In cases of addiction or significant negative impact on mental health, psychological interventions such as cognitive-behavioral therapy can be helpful. Yes, psychological interventions are fundamental in cases of addiction or significant negative impact on mental health. Cognitive-behavioral therapy (CBT) is one of the most widely used and supported therapeutic modalities to address a variety of psychological issues, including addiction disorders.

Cognitive-behavioral therapy focuses on identifying and changing dysfunctional patterns of thought and behavior that contribute to mental health problems. In the context of addiction, CBT can help individuals understand and modify the beliefs and attitudes that sustain their addictive behavior. Some common goals of CBT in treating addiction include:

Helping the person recognize and change negative or irrational thoughts that may contribute to addiction.

Teaching effective strategies for managing stress, anxiety, or other emotions that may trigger addictive behavior.

Working on identifying risk factors and developing strategies to prevent relapse into substance use or addictive behaviors.

Assisting the person in changing behavior patterns that contribute to addiction and promoting healthy behaviors.

In addition to cognitive-behavioral therapy, there are other forms of psychological intervention that can also be helpful, depending on the specific nature of the addiction and individual needs. These may include motivational therapy, group therapy, family therapy, among others.

It's important to note that the choice of psychological intervention may vary depending on the individual and the severity of the addiction. In many cases, a combination of therapeutic approaches and, in some instances, collaboration with medications, can be the most effective strategy to address addiction and promote long-term recovery. Active engagement and commitment from the individual are key to the success of any psychological intervention.

Family involvement in understanding and setting healthy limits can be crucial in mitigating risks associated with video games. Absolutely, family support plays a crucial role in addressing issues related to video games and establishing healthy habits in their use. Here are some ways in which family support can be beneficial:

Fostering an environment where family members feel comfortable sharing their concerns about video game use. Open communication can help better understand each person's motivations and experiences.

Working together to set healthy time limits for video game use. These limits can help prevent excessive use and ensure a proper balance between online and offline activities.

Actively engaging in activities related to video games, such as playing together, understanding the games family members enjoy, and discussing digital experiences openly and without judgment.

Informing the family about potential risks associated with excessive video game use, such as social isolation, lack of physical activity, and effects on mental health. Shared awareness can facilitate the adoption of healthier habits.

If signs of addiction or significant negative impacts on mental health are observed, seeking help from a mental health professional may be crucial. Therapists specialized in addictions or mental health can offer guidance and support.

Encouraging and participating in screen-free activities, such as sports, outdoor activities, reading, or board games. Diversifying activities can contribute to a more balanced lifestyle.

It's essential to remember that the key is to find a healthy balance in the use of video games and other activities. Active family involvement in understanding, supporting, and setting limits can significantly contribute to mitigating the risks associated with excessive video game use and fostering a more balanced and healthy family environment.

18.Data Tracking and Privacy

Anxiety associated with the massive collection of data and lack of online privacy is an increasingly common phenomenon in the digital era. Here are some reasons why people may experience anxiety regarding this issue:

The massive collection of data often means that a large amount of personal information is in the hands of third parties, such as companies and online platforms. The feeling of losing control over one's personal information can lead to anxiety.

The sense of losing control over one's personal information, derived from the massive data collection by third parties, is a legitimate concern in the digital age. Here are some reasons why this loss of control can cause anxiety:

In many cases, data collection practices by companies and online platforms are not entirely transparent. Users may not be fully informed about what data is being collected, how it is used, and with whom it is shared.

The collected information can be used in ways that users have not authorized or anticipated. This includes selling data to third parties, creating detailed profiles, and personalizing advertising, which can cause unease.

The frequency of data breaches and cyberattacks increases concerns about the security of personal information. The possibility of data falling into the wrong hands or being subject to unauthorized access can be stressful.

The personal information collected can be used to manipulate users' perceptions and behavior. This can cause anxiety, as users may feel that information is being used to influence decisions and opinions unethically.

Data collection can create a sense of being under constant online surveillance. This sense of being watched can affect personal freedom and privacy, generating anxiety about the loss of anonymity.

To address this anxiety and regain some degree of control over personal information, users may consider taking the following measures:

Utilize privacy settings on platforms and devices to control who has access to personal information and what permissions are granted.

Familiarize themselves with the privacy policies of the platforms and services used. Understanding how data is collected, used, and shared can provide clarity.

Be selective when sharing personal information online and avoid providing unnecessary details in profiles and forms.

Use tools such as VPNs, tracking blockers, and privacy extensions in browsers to enhance online security and anonymity.

Stay informed about online privacy and security practices, as well as updates to the privacy policies of the platforms used.

By adopting conscious practices and proactive measures, users can mitigate the sense of losing control and reduce anxiety associated with the massive collection of data online.

The increasing frequency of data breaches and cyberattacks can heighten concerns about the security of personal information. The fear that sensitive information may fall into the wrong hands can contribute to anxiety.

Concerns about online security are understandable and well-founded, especially given the growing frequency of data breaches and cyberattacks. The security of personal information has become a priority for many users, and anxiety related to this issue may arise for various reasons:

The existence of previous cases of large-scale data breaches, where large amounts of personal information have been compromised, contributes to widespread concern about online security.

The fear of potential consequences of a data breach, such as identity theft, financial fraud, identity impersonation, and other cybercrimes, can generate anxiety in users.

The sense of lack of control over the security of personal information, especially when relying on third parties to safeguard that data, can be a significant source of concern.

The concern that stolen personal information may be misused or sold on the black market increases anxiety about online security.

To address these concerns and reduce anxiety associated with online security, users may consider taking the following measures:

Use strong and unique passwords for each online account to reduce the risk of unauthorized access.

Enable two-factor authentication whenever possible to add an additional layer of security.

Keep software, including operating systems and applications, up to date to benefit from the latest security patches.

Be vigilant against phishing attempts and avoid clicking on links or providing personal information in suspicious emails or messages.

Review and adjust privacy settings on online platforms to limit the amount of publicly shared information.

Regularly monitor online accounts for unusual activity and frequently review financial statements.

Consider using VPNs to encrypt internet connections and protect transmitted information online.

Stay informed about the latest threats and best practices for online security.

Taking a proactive approach to online security and following sound security practices can help reduce anxiety and enhance the protection of personal information in the digital environment.

Data collection is often used to create detailed user profiles for the purpose of customizing advertising and services. The feeling of being observed and the lack of anonymity can generate unease and anxiety.

Creating detailed user profiles for the customization of advertising and services is a common practice in the massive collection of online data. However, this practice can cause unease and anxiety in users for several reasons:

The idea that detailed data about online and offline behavior is being collected can create the feeling of being constantly observed. This can make users feel vulnerable and concerned about the lack of privacy.

Creating detailed profiles implies that third parties have access to intimate information about preferences, behaviors, and habits. The lack of control over how this information is used can contribute to anxiety.

If detailed profiles are used to intrusively customize ads, displaying highly specific and relevant content to the user, this can create a sense of privacy invasion and contribute to anxiety.

The detailed information collected can be used to influence users' decisions and opinions in ways that may not be apparent to them. Concerns about manipulation based on detailed profiles can increase anxiety.

To manage anxiety associated with the creation of detailed profiles, users may consider the following actions:

Review and adjust privacy settings on online platforms to limit the amount of data collected and shared.

Where possible, use "opt-out" options to avoid customization based on detailed profiles.

Educate oneself about how privacy settings work on the platforms used and make informed decisions about what information to share.

Use tools and browser extensions that block or limit online tracking.

Be selective when providing personal information online and limit the disclosure of details that could contribute to detailed profiles.

Understand that personalized advertising is a common practice and be vigilant against advertising strategies without succumbing to manipulation.

Conscious management of online privacy and informed decision-making can help reduce anxiety associated with the creation of detailed profiles and the customization of online services.

Targeted advertising based on data collection can make people feel invaded in their privacy. The feeling that platforms know too much about personal preferences can be intrusive.

Personalized advertising based on data collection can indeed generate a feeling of privacy invasion among users. Here are some reasons why this may occur:

Personalized advertising relies on detailed knowledge of users' preferences, behaviors, and habits. The feeling that platforms know too much about personal life can be intrusive and cause unease.

Users may often feel that they have not given explicit consent for their data to be used for advertising purposes, especially when data collection and use are not clearly communicated.

When personalized advertising becomes too accurate, users may perceive that ads appear at inconvenient times and in contexts they consider invasive.

Personalized advertising may be designed to influence users' decisions and behaviors in ways that may seem manipulative. This can raise concerns about the lack of control over external influences.

To manage the feeling of privacy invasion associated with personalized advertising, users may consider the following actions:

Review and adjust privacy settings on online platforms to limit the amount of data collected and shared for advertising purposes.

Use "opt-out" options when available to avoid personalized ad customization based on personal data.

Educate oneself about how online advertising practices work and understand available options for controlling ad personalization.

Use tools and browser extensions designed to block or limit online tracking and data collection for advertising.

Be aware that personalized advertising is a common online practice and develop a critical understanding of how personal information is used for advertising purposes.

Active management of privacy settings and informed decision-making can help users mitigate the feeling of privacy invasion associated with personalized advertising based on data collection.

Concerns about how collected data is used and whether it can be misused, such as being sold to third parties or manipulated for unethical purposes, can contribute to anxiety.

Concerns about the misuse of data are a legitimate and understandable concern in the digital age. Lack of transparency and fear that collected information may be misused can significantly contribute to user anxiety. Here are some reasons why this concern may arise:

When companies are not transparent about how collected data is used, users may feel uncertainty and concern about the fate of their personal information.

The possibility that collected data may be sold to third parties for commercial purposes can generate unease. Users may feel they are losing control over information and its ultimate destination.

Concern that collected information may be used to manipulate decisions and opinions for unethical purposes can increase anxiety. This may include manipulation of political preferences, purchasing behaviors, and other actions.

Anxiety may arise when users feel they have lost control over their data and have not given clear and specific consent for its use.

To address anxiety associated with the misuse of data, users may consider the following actions:

Read and understand the privacy policies of the platforms and services used to have a clear idea of how data is collected, used, and shared.

Adjust privacy settings on online platforms to limit the amount of data shared and with whom it is shared. Use "opt-out" options whenever possible to prevent personal information from being used for concerning purposes. Stay informed about online data collection practices and understand how companies handle user information. Use tools and browser extensions that block or limit online data collection. Participate in initiatives and support legal measures and regulations that promote online privacy protection. The combination of informed decision-making, active privacy settings, and engagement in data protection dialogue can help reduce anxiety related to potential misuse of personal information.

The feeling of being constantly tracked online, whether through cookies, location tracking, or behavior analysis, can create a sense of constant surveillance, contributing to anxiety.

The feeling of being constantly tracked online can indeed contribute to a sense of constant surveillance and generate anxiety. Various online tracking methods, such as the use of cookies, location tracking, and behavior analysis, can contribute to this concern. Here are some reasons why this can generate anxiety:

The persistence of online tracking can make users feel identified and not anonymous while browsing the web. The loss of anonymity can raise concerns about privacy and the possibility of unwanted tracking.

While online service personalization can be helpful, the feeling that every online action is being tracked to customize ads and content can seem invasive, generating anxiety about the loss of privacy.

Location tracking can be particularly intrusive, as it reveals a person's physical location in real time. Concerns about who has access to this information and how it is used can contribute to anxiety.

The perception of a lack of control over online tracking can generate anxiety. Users may feel they are being tracked without their knowledge or explicit consent.

To manage anxiety associated with constant online tracking, users may consider the following actions:

Review and adjust privacy settings in browsers and devices to limit online tracking. This may include regularly deleting cookies and configuring privacy preferences.

Use VPNs: Use virtual private networks (VPNs) to encrypt the internet connection and protect information transmitted online, helping to preserve anonymity.

Employ browser extensions or applications that block or limit online tracking, including cookie blockers and privacy protection tools.

Review and adjust privacy settings in mobile apps that use location tracking or other forms of tracking.

Stay informed about online tracking practices and how specific platforms and services handle user information.

Explore the use of browsers specifically designed to protect privacy, which often include built-in features to limit tracking.

Active management of privacy settings and the use of tools that limit online tracking can help users feel more in control and reduce anxiety associated with constant surveillance in the digital environment.

To manage anxiety associated with massive data collection and lack of online privacy, some strategies may include:

Leveraging privacy settings available on platforms and devices to control who has access to personal information.

Privacy settings are an essential tool for users to control who has access to their personal information on online platforms and devices. Here are some general steps users can take to leverage these settings and strengthen their online privacy:

When creating an account on a new platform or setting up a device, it's important to review privacy settings from the start. Many platforms offer specific options during the initial setup process.

Access the account settings section on the platform and review available privacy options. This may include settings related to profile visibility, personal information, and privacy preferences.

On social media platforms and online services, review and adjust privacy settings for profile management and content visibility. There may be options to control who can see posts, photos, friends, etc.

Review and adjust privacy settings related to personal information, such as email address, phone number, and date of birth. Limit access to this information only to trusted individuals.

Set up additional security measures, such as two-factor authentication (2FA), to strengthen account protection and prevent unauthorized access.

Evaluate and manage settings related to location and tracking on mobile devices. Some apps and services allow disabling location tracking or limiting its use.

Review settings related to personalized advertising and online activity tracking. Many platforms offer options to control or limit ad personalization based on personal data.

Configure cookie preferences in web browsers to limit tracking. Some browsers offer private browsing modes that block certain forms of tracking.

Conduct regular reviews of privacy settings, especially after platform updates or changes in privacy policies. Ensure that settings align with current preferences.

Stay informed about the privacy features of the platforms and devices used. Updates and new features often introduce additional configuration options.

Each platform and device may have specific privacy settings, so it's important to carefully review available options in each case. Active management of privacy settings allows users to have greater control over who accesses their personal information and contributes to a safer and more personalized online experience according to their preferences.

Getting informed about data collection practices and understanding how to protect privacy online can empower individuals and reduce anxiety.

Digital education plays a crucial role in empowering individuals to understand and protect their online privacy. Here are some ways in which digital education can help reduce anxiety associated with data collection and enhance online security:

Digital education provides people with information about how online data collection practices work. This includes understanding what data is collected, how it is used, and who has access to it. Understanding these practices can help demystify the process and reduce anxiety.

Learning about privacy settings available on different online platforms and services allows people to make informed decisions about how they want to share their personal information. Knowing how to adjust these settings can significantly improve control over privacy.

Digital education can provide information about the risks associated with excessive disclosure of personal data online. Understanding potential consequences, such as loss of privacy, identity theft, and exposure to unwanted advertising, can motivate people to be more cautious.

Knowing and learning to use privacy tools, such as tracking blockers, privacy extensions in browsers, and virtual private networks (VPNs), can enhance online security and provide an additional layer of protection.

Digital education can address topics related to password security, two-factor authentication (2FA), and other practices to protect online accounts. Knowing and applying good security practices contributes to preventing unauthorized access.

Developing digital discernment skills involves learning to critically evaluate information online, identify potential threats, and recognize phishing tactics. This helps avoid situations that could compromise privacy.

Being informed about privacy laws and user rights online. Knowing legal protections and individual rights can strengthen people's position in protecting their privacy.

Digital education also includes staying aware of updates in the privacy policies of regularly used online platforms. Policy changes can affect how personal data is handled.

By cultivating a strong level of digital education, individuals can make more informed decisions and feel more empowered to protect their privacy online. Reducing anxiety associated with data collection goes hand in hand with acquiring knowledge and skills that enable more conscious and secure use of technology.

Using tools such as VPNs (virtual private networks), tracking blockers, and privacy settings in the browser to enhance online security.

The use of privacy protection tools is an effective strategy to increase online security and reduce exposure to unwanted data collection. Here are some common tools and practices that people can employ to protect their privacy online:

A VPN encrypts the internet connection, hiding the user's IP address and masking their physical location. It provides anonymity while browsing the web, protects against location tracking, and encrypts transmitted information, making it more secure.

Extensions like uBlock Origin or Privacy Badger block trackers and scripts used by websites and advertisers to collect data. They reduce the amount of data collected during browsing and improve privacy by preventing third-party tracking.

Modern browsers offer specific privacy settings. This includes options to manage cookies, block unwanted content, and adjust privacy settings. It allows users to customize their online experience and limit tracking by adjusting settings according to their preferences.

Extensions like HTTPS Everywhere or DuckDuckGo Privacy Essentials enhance privacy by ensuring secure connections and blocking unwanted elements. They increase online security by encrypting connections and preventing the execution of unwanted scripts.

Using privacy-focused search engines, such as DuckDuckGo or Startpage, that do not track user searches. It protects privacy by avoiding the collection of personalized search data by conventional search engines.

Tools like LastPass or 1Password help manage passwords securely and encourage the use of strong and unique passwords. It improves security by preventing password reuse and protects accounts against unauthorized access.

Keeping the operating system and applications updated ensures that security patches are applied, and vulnerabilities are fixed. It enhances the overall security of the device and reduces the risk of attacks based on known vulnerabilities.

Configuring the browser to block third-party cookies, limiting websites' ability to track online activity. It reduces tracking and data collection by websites and advertisers.

By combining these tools and practices, individuals can significantly strengthen their online privacy and reduce anxiety associated with massive data collection. It's important to note that the choice of tools and practices should align with individual needs and comfort levels.

Critically evaluating the use of platforms that collect large amounts of data and considering limiting involvement in those causing greater concern.

Limiting the use of certain platforms that collect large amounts of data is a valid strategy for users seeking to reduce exposure and anxiety associated with massive online data collection. Here are some steps and considerations to keep in mind when critically evaluating the use of platforms and considering limiting involvement:

Carefully read the privacy policies of the platforms used to understand how user data is collected, used, and shared.

Assess a platform's transparency regarding its data handling and the clarity of its privacy policies.

Critically evaluate whether the platform is essential to meet personal or professional needs.

Weigh the benefits derived from using the platform against the perceived risks associated with data collection.

Look for alternatives that offer similar services but with stronger privacy protection approaches.

Consider using open-source platforms or those that allow greater control over privacy and settings.

If deciding to continue using a platform, review and adjust available privacy settings to limit the amount of shared data.

Limit the amount of personal information shared in profiles and online interactions. Avoid sharing especially sensitive information if not strictly necessary. Consider periods of disconnection or regular digital breaks to reduce constant exposure and the accumulation of online data. Stay informed about online data collection practices and understand how they impact privacy. Develop digital discernment skills to identify and avoid situations that may compromise privacy. Consider supporting and participating in initiatives and platforms that adopt ethical practices regarding privacy and data security.

Limiting the use of platforms that raise concerns in terms of data collection can be an informed and empowered decision. By critically evaluating platforms and adopting conscious online practices, users can exert greater control over their privacy and reduce anxiety associated with constant exposure to data collection. Managing anxiety related to online privacy may require a holistic approach that includes changes in online behavior, education on digital privacy, and the adoption of tools that protect personal information.

19.Technology and Personal Authenticity

The relationship between technology and personal authenticity is complex and can have various impacts on how people express their true selves. Social media and online platforms can facilitate connections with friends, family, and like-minded communities, allowing individuals to share authentic aspects of their lives.

Certainly, connecting through social networks and online platforms can have several positive aspects in terms of facilitating authentic expression and connection among individuals. Here are some reasons why social media can be beneficial in this regard:

Social networks enable people to connect with friends, family, and communities worldwide, expanding the possibilities of establishing meaningful relationships with those who share similar interests, regardless of physical location.

Online platforms provide a space for individuals to authentically share their life experiences, achievements, and challenges, strengthening emotional bonds and creating a sense of support and understanding.

Social media allows the formation and participation in communities of shared interests, where individuals can find groups reflecting their values, passions, or identities, facilitating authentic expression in a mutually understanding environment.

Online platforms offer a space for the expression of diverse identities, opinions, and perspectives, promoting diversity and inclusion by allowing authentic voices to be heard and respected.

Social networks provide opportunities for emotional and social support, especially beneficial in challenging times, as they enable the sharing of experiences and receiving support from the online community.

Through constant online communication, people can maintain and strengthen relationships even in situations where physical distance would be a barrier, contributing to the building of meaningful connections over time.

Online platforms offer spaces for creative self-expression, whether through art, writing, photography, or other mediums, allowing individuals to authentically share their talents and passions.

Online exposure can provide professional, educational, or creative opportunities by showcasing authentic achievements and skills, potentially opening doors to new prospects.

Social media and online platforms have the potential to enrich people's social and emotional lives by facilitating authentic connections with others. The key is to use these tools consciously and balancedly to harness their positive benefits.

Digital tools and creative platforms provide people with opportunities to express their creativity uniquely, whether through music, art, writing, or digital content creation. Creative self-expression through digital tools and online platforms offers numerous positive opportunities:

Online platforms allow creativity to reach global audiences, as artworks, music, or digital content can be shared and appreciated worldwide, providing exposure and recognition opportunities.

Digital tools offer a wide variety of creative mediums, from digital illustration to music production and multimedia content creation, allowing individuals to explore and express their creativity in diverse ways.

Online platforms facilitate remote creative collaboration, enabling artists to work together on projects despite geographical separation, fostering diversity of perspectives and styles.

The availability of online tutorials and educational resources allows people to learn new creative skills, stimulating growth and evolution in creative expression as knowledge is acquired and techniques are perfected.

Digital platforms remove traditional barriers to entry in creative fields, democratizing artistic expression as anyone with access to digital tools can explore and share their creativity.

Participation in online communities allows for constructive feedback and support from other creatives, fostering growth and continuous improvement through positive interactions and collaborations.

Digital tools provide creators with the ability to experiment with their art uniquely and originally, facilitating authentic expression of personal creative vision without the limitations of traditional media.

Digital platforms enable self-publishing and distribution of creative works, allowing artists to share and sell their work directly to the public, having greater control over their creative narrative and career.

Digital tools allow the creation of interactive and participatory experiences, engaging the audience more directly and creating a deeper connection between the creator and their audience.

In summary, creative self-expression through digital tools and online platforms has transformed the way people explore, share, and experiment with their creativity. These positive opportunities have contributed to the diversification and democratization of artistic and creative expression.

Technology provides access to information and resources that can support personal development and self-discovery, allowing individuals to explore and better understand their authenticity. Technology enables access to up-to-date and relevant information in real-time, facilitating the search for educational resources, expert advice, and data that support personal development. The availability of online learning platforms offers people the opportunity to acquire new skills, explore interests, and enhance their knowledge in specific areas, contributing to personal growth.

Technology facilitates the creation and participation in online communities that share similar interests or challenges, providing a space for the exchange of experiences and mutual support. There are applications and online platforms dedicated to personal development, offering resources such as e-books, podcasts, courses, and meditations. These resources can be accessible from anywhere, providing flexibility to users.

Technology enables connections with professionals in various fields through online services, virtual consultations, and professional networks, facilitating the search for guidance and specialized support. Applications and online tools allow people to conduct self-assessments and track their progress in personal goals, contributing to self-awareness and authenticity by recognizing achievements and areas for improvement.

Technology has facilitated access to online therapy services and psychological support applications, offering accessible options for those seeking assistance in their mental and emotional well-being. Online platforms provide information on health and well-being, from dietary advice to exercise routines, enabling individuals to make informed decisions to improve their quality of life. Online communities and digital resources offer support to those facing personal challenges, such as loss, addiction, or stress management.

Digital connectivity can mitigate isolation and provide valuable resources. Technology allows exploration and understanding of diverse cultural perspectives and life experiences, contributing to open-mindedness, empathy, and understanding of authenticity in diverse cultural contexts.

Access to information and support through technology plays a crucial role in personal development and self-discovery, facilitating learning, connecting with like-minded communities, and providing resources that support authenticity and personal growth. Online communication allows connections with people worldwide, expanding perspectives and offering opportunities to explore different identities and forms of authenticity.

Online communication effectively expands perspectives and offers opportunities to explore different identities and forms of authenticity. Online communication provides the opportunity to interact with people from diverse cultures and backgrounds, enriching cultural understanding, fostering open-mindedness, and promoting acceptance of various forms of authenticity. The ability to connect with people from different parts of the world facilitates the exchange of perspectives and life experiences, contributing to a deeper understanding of the various ways authenticity can manifest.

Online communication offers a space for individuals to explore and express different facets of their identity, providing a safe medium to discover and share authentic aspects of themselves. Online platforms can be safe spaces for members of minorities and marginalized groups to find support and understanding, facilitating connections with people who share similar experiences and strengthening collective authenticity. Social media and online forums enable participation in global communities focused on specific interests, where individuals can feel free to express their authenticity in an environment of mutual understanding.

Online communication facilitates collaboration on creative, professional, or social projects at the international level, expanding opportunities to work with people from diverse cultures and lifestyles. Online interaction provides opportunities for intercultural learning by exposing people to ideas, traditions, and values from around the world, promoting acceptance and appreciation for the diversity of forms of authenticity.

Online communication allows maintaining meaningful connections with friends, family, or loved ones who are geographically distant, facilitating the continuity of valuable relationships despite physical barriers.

Online platforms are used for organizing global movements and activism, uniting people with common goals and allowing authenticity to be expressed in the fight for important causes. Online communication can contribute to the development of social skills by providing opportunities to interact with a variety of people, especially beneficial for those facing social barriers in physical environments. In summary, online communication plays an essential role by providing a global space where people can connect, learn, and express their authenticity in diverse ways. This global access contributes to the richness of the human experience and promotes a deeper understanding of the various forms of being authentic.

The prevalence of filters and image editing tools on social media can distort real physical appearances, creating unrealistic expectations and pressures to conform to certain beauty standards. The widespread use of image filters and editing tools on social media has significant negative impacts as it can distort real physical appearances and create unrealistic expectations.

The widespread use of filters and editing tools contributes to the creation of unrealistic beauty standards. Edited images may present an idealized and perfected version of physical appearance, generating pressures for individuals to conform to these unattainable standards.

The constant exposure to edited images can distort the perception of one's own body image. People may experience dissatisfaction with their real appearance when comparing themselves to artificially enhanced images.

Exposure to edited images can have a negative impact on self-esteem. Individuals may feel insecure or dissatisfied with their natural appearance when comparing it to retouched images they see online.

The prevalence of edited images can normalize the idea that certain unattainable beauty standards are the norm. This contributes to the construction of a culture that values visual perfection over authenticity.

In the realm of personal relationships, the excessive use of filters can create unrealistic expectations about people's appearances. This can have implications for how relationships are established and maintained, based on images that do not reflect reality.

The pressure to meet unrealistic beauty standards can contribute to mental health issues such as anxiety and depression. Constantly comparing oneself to retouched images can lead to a sense of inadequacy and negative self-evaluation.

The culture of image editing can foster body insecurity by suggesting that natural imperfections should be eliminated or hidden. This can lead to a negative relationship with one's own body.

Image editing can lead to a falseness in the representation of reality. This undermines online authenticity by presenting idealized and perfected versions that do not reflect the real diversity and authenticity of individuals.

The need to maintain a perpetual image of perfection can generate stress and anxiety. People may feel pressured to always appear flawless, which can be exhausting and unrealistic.

Excessive image editing can affect the construction of personal identity. Individuals may come to identify more with their edited images than with their natural appearance, which can have consequences for authenticity and self-acceptance.

The prevalence of filters and image editing tools on social media can have significant negative consequences on people's perception of appearance, self-esteem, and mental health by perpetuating unrealistic beauty standards.

Social media often encourages social comparison, leading to the suppression of authenticity to meet perceived expectations of others.

Social media-induced social comparison can have significant negative effects, as it can lead to the suppression of authenticity and generate pressures to meet perceived expectations.

Social media often presents idealized versions of people's lives, contributing to the perception of unattainable standards in terms of appearance, achievements, and lifestyle. This can create pressures to conform to these unrealistic expectations.

Constant comparison with the seemingly perfect lives of others on social media can trigger negative self-evaluation. People may feel that their own lives are not as successful or satisfying as those of their online contacts.

Social comparison can have a negative impact on self-esteem. Individuals may experience feelings of inadequacy and develop a negative view of themselves by constantly measuring themselves against the achievements and appearances of others.

The pressure to meet certain social expectations can lead to the suppression of authenticity. People may feel the need to present an idealized version of themselves online, hiding genuine aspects for fear of judgment or comparison.

Constant comparison on social media can contribute to social anxiety. People may feel constantly evaluated by their peers online, increasing stress and perceived social pressure.

Social comparison can lead to an excessive focus on seeking external approval. Individuals may base their sense of worth on the number of "likes," comments, or validation they receive on their posts, rather than cultivating intrinsic self-esteem.

Constant comparison on social media can contribute to mental health issues, such as depression and anxiety. The gap between the perceived online life and reality can cause significant psychological distress.

Social comparison can lead to a lack of authenticity in online interactions. People may feel the need to present only positive aspects of their lives, contributing to a biased and inauthentic representation.

Constant comparison can affect personal relationships, as individuals may feel competitive or envious. This can lead to tensions and misunderstandings instead of genuine connections.

Constant comparison can lead to the redefinition of personal values to fit online expectations. People may deviate from what they truly value to conform to what they perceive as desired by others.

Overall, social comparison on social media can have negative consequences for personal perception and the quality of online interactions. Fostering digital awareness and promoting a culture of acceptance and authenticity can help mitigate these adverse effects.

Online anonymity can lead to the creation of false identities or the expression of behaviors that differ from one's authentic personality, making genuine authenticity difficult.

The creation of false identities online, facilitated by internet anonymity, can have various negative effects, making genuine and authentic expression challenging. Here are some negative aspects associated with adopting false identities online:

The adoption of false identities can lead to deception and a lack of trust in online interactions. People may feel disappointed or betrayed upon discovering that the online representation does not align with reality.

The presence of false identities complicates the building of authentic and genuine relationships online. Lack of transparency can affect the quality of connections and undermine mutual trust.

Online anonymity facilitates cyberbullying and abuse, as individuals may feel protected by hiding their true identity. This can have severe consequences for the mental and emotional health of victims.

False identities can contribute to less authentic communication. People may feel the need to present altered versions of themselves, limiting honesty and openness in online interactions.

Creating and maintaining false identities can have a negative impact on the mental health of those adopting these identities. The disconnect between online life and reality can cause internal conflict and stress.

Adopting false identities can lead to a disconnection from reality. Individuals may lose authenticity by immersing themselves in fictional roles, affecting their perception and understanding of the real world.

False identities can contribute to the spread of misinformation online. The lack of verification of the authenticity of sources can lead to the dissemination of fake news and deceptive content.

The presence of false identities can complicate conflict resolution online. Lack of personal accountability makes it difficult to address and resolve disputes effectively.

The prevalence of false identities can contribute to the devaluation of online authenticity. People may begin to assume that most online representations are fictional, undermining trust and genuine connection.

In some cases, adopting false identities can have legal consequences, especially if used for fraudulent or criminal purposes. Lack of accountability may lead to legal actions in cases of cybercrimes.

The presence of false identities online poses significant challenges to authenticity and the quality of interactions in the digital environment. Fostering transparency and responsibility online is essential to mitigate these negative effects.

Overexposure to online information can overwhelm individuals, making it difficult to identify and express what is truly meaningful to them.

Online information overload can have negative impacts, making it difficult for individuals to identify and express what is truly meaningful to them. Here are some negative aspects associated with this phenomenon:

Information overload can lead to cognitive overload, overwhelming an individual's mental processing capacity. This makes it difficult to identify and focus on what truly matters to each person.

The abundance of information can make it challenging to prioritize what is genuinely meaningful. People may feel lost amid the data overload, leading to suboptimal decision-making and a lack of focus.

Constant exposure to large amounts of information can generate fatigue. People may feel exhausted and disconnected, affecting their ability to identify and express their true needs and desires.

Information overload can distort priorities, causing individuals to pay attention to what is popular rather than what is truly meaningful to them. This can influence decision-making and authentic expression.

Overexposure to online information can contribute to a lack of personal connection. Interactions become more superficial, and people may struggle to express their authentic experiences and feelings.

Information saturation can stifle creativity by limiting people's ability to explore new ideas and perspectives. Authentic expression is often compromised when minds are overwhelmed by information.

The abundance of choices can create choice anxiety, where people feel overwhelmed by the amount of information available. This can result in decision paralysis and an inability to express authentic preferences.

Information overload leaves little time for reflection. The ability to consider and authentically express thoughts and feelings may be limited when individuals are constantly bombarded with new information.

The abundance of online information facilitates constant comparison with others. People may feel pressured to conform to external standards rather than authentically expressing what they truly value.

Information overload can disrupt the process of self-discovery. Individuals may be so busy processing external information that they struggle to explore and understand their own authentic needs and desires.

Online information overload can have negative consequences by overwhelming individuals' ability to identify and express what is truly meaningful to them. Conscious information management and prioritizing authenticity are key elements in counteracting these negative impacts.

Concerns about online privacy can lead to self-censorship and limitations in sharing authentic experiences for fear of unwanted exposure.

Online privacy risks can have significant negative effects, contributing to self-censorship and limiting the sharing of authentic experiences. Here are some negative aspects associated with these privacy risks:

Concerns about privacy can lead to self-censorship, where individuals avoid sharing authentic experiences for fear of unwanted exposure. This limits genuine expression and connection with others online.

Privacy concerns may result in restricting personal expression online. People may avoid sharing authentic aspects of their lives due to concerns about privacy invasion or misuse of information.

Fear of privacy risks may lead to limitations in participating in online communities. Individuals may avoid contributing to authentic discussions or sharing experiences due to concerns about the security of their personal information.

Concerns about privacy can create mistrust in online platforms. This can lead to a decrease in participation and authentic expression, as individuals doubt the security of their data.

The fear of privacy risks can contribute to a culture of secrecy and caution online. People may be hesitant to share genuine experiences, affecting the depth and authenticity of online interactions.

Worries about privacy can hinder the building of trust in online relationships. Individuals may be reluctant to open up authentically, fearing potential breaches of their privacy.

Privacy concerns can lead to a reluctance to explore personal aspects of one's life online. This may limit the richness of online interactions and the expression of authentic experiences.

The perception of privacy risks can contribute to a reduction in the authenticity of online communication. People may feel the need to guard their personal information, leading to less open and genuine conversations.

The fear of privacy risks may result in avoiding online self-disclosure. Individuals might refrain from sharing authentic thoughts and feelings, limiting the depth of their online connections.

Privacy concerns can contribute to a hesitancy to express oneself authentically online. Striking a balance between sharing personal experiences and safeguarding privacy is crucial for maintaining online authenticity.

In conclusion, privacy concerns in the online environment can lead to negative consequences such as self-censorship and limitations in the sharing of authentic experiences. Striving for a balance between privacy and authenticity is essential for fostering genuine online interactions.

The constant concern for online privacy can have an impact on mental health. Individuals may experience anxiety and stress, fearing the potential unauthorized exposure of personal information.

Privacy concerns may result in less transparency in online interactions. People may choose to present more limited or selective versions of themselves, affecting the authenticity of digital relationships.

Privacy concerns intensify with the increasing frequency of data breaches. The threat of personal information being exposed may lead to self-censorship and a decrease in online participation.

Privacy concerns can inhibit the exploration of personal interests online. Individuals may avoid seeking information on specific topics or participating in communities that could reveal more personal aspects of their lives.

Fear of privacy risks can result in reduced participation in online activities. People may refrain from contributing to blogs, forums, or social networks due to fear of unwanted exposure.

Distrust in data protection measures can lead to limitations in information sharing. People may hesitate to provide personal data even when necessary for certain online activities.

In summary, online privacy risks can have negative consequences by limiting authentic expression and participation in digital communities. Addressing these concerns requires a balance between privacy protection and facilitating an online environment where people feel safe to express themselves genuinely.

Promoting digital awareness so that individuals understand how technology affects their personal expression and adopt practices that encourage authenticity and well-being is crucial.

Digital awareness is essential in the modern era for people to understand how technology affects their personal expression and adopt practices that encourage authenticity and well-being. Here are some important points related to digital awareness:

Foster awareness of the importance of online privacy. This includes understanding how personal information is collected, used, and shared in the digital environment.

Provide education on the risks and benefits associated with the use of technology. This allows people to make informed decisions about how to use digital tools in a way that aligns with their values and needs.

Promote the idea that online authenticity is valuable, and people should not feel pressured to conform to unrealistic standards. Authentic expression contributes to the construction of healthier digital communities.

Develop the ability to discern information manipulation online, including understanding fake news, misinformation, and the influence of algorithms on content presentation.

Recognize that online representation is selective and often does not reflect the entirety of reality. Understand that people present edited and filtered versions of their lives on digital platforms.

Teach the effective use of privacy settings available on social media platforms and devices. This allows people to have greater control over who has access to their personal information.

Encourage healthy practices on social media, such as setting time limits, promoting positive interactions, and identifying and managing social comparison.

Providing digital security training to help people protect their personal information and prevent threats such as phishing, identity theft, and other forms of cybercrime.

Developing critical digital media skills to evaluate the reliability of online information and understand how digital platforms can influence perception and behavior.

Promoting digital empathy to cultivate a deeper understanding of others' experiences online. This contributes to the building of more inclusive and respectful online communities.

Digital awareness is essential to empower individuals in their interaction with technology and ensure they use digital tools in a way that fosters authenticity, well-being, and online security.

Encouraging conscious use of social media, recognizing the benefits but also setting healthy limits to avoid social comparison and pressure.

Promoting conscious use of social media is crucial to ensure that people harness the benefits of these platforms while setting healthy limits. Here are important points related to conscious use of social media:

Educate people about the benefits and risks associated with the use of social media. This includes promoting meaningful connections and awareness of potential negative effects, such as social comparison and social pressure.

Encourage the practice of setting time limits on social media use. This helps prevent overexposure, digital fatigue, and unnecessary time loss.

Incentivize positive and constructive interactions on social media. This involves promoting encouraging comments, supporting others' experiences, and contributing to a healthy online environment.

Develop awareness of social comparison on social media. People should understand that online representations can be selective and do not reflect the entirety of reality.

Encourage authenticity in posting content on social media. This helps build genuine connections and counteracts the pressure to conform to unattainable standards.

Provide education on privacy settings available on social media platforms. People should understand how to adjust these settings to control who has access to their information.

Encourage the practice of occasional disconnection from social media. These digital breaks can help reduce dependency and improve emotional well-being.

Create awareness of how algorithms affect content presentation on social media. Understanding that platforms select and display content based on their algorithms can shift perceptions of reality.

Shift the focus toward building meaningful connections rather than accumulating followers or "likes." The quality of interactions is more important than quantity.

Equip individuals with skills to cope with online social pressure. This includes the ability to set boundaries, say no to unrealistic expectations, and focus on their own well-being.

By promoting conscious use of social media, we contribute to a healthier and more balanced digital environment. Education, awareness, and the promotion of positive practices are key to reaping the benefits of social media without compromising individuals' mental and emotional health.

Promoting digital authenticity involves encouraging people to share genuine experiences, be honest about their online lives, and create environments where authenticity is valued.

Promoting digital authenticity is essential for fostering healthy online environments and meaningful connections. Here are important points related to promoting digital authenticity:

Encourage people to share genuine and authentic experiences online. This creates deeper connections and contributes to a digital environment where each person feels valued for their uniqueness.

Promote honesty in online representation. Encourage people to present authentic versions of themselves, avoiding the pressure to conform to unrealistic standards or create a digital facade.

Establish online communities that support authenticity. Provide spaces where people feel safe to share their experiences without fear of judgment, thus fostering genuine expression.

Incentivize positive vulnerability, where people share their challenges and achievements openly and constructively. This strengthens connections and reduces the stigma associated with personal experiences.

Foster diversity of narratives online. Recognize and celebrate different perspectives and experiences, contributing to a more inclusive and enriching digital landscape.

Help people understand how genuine expression contributes to personal well-being and the building of meaningful relationships.

Demystify the idea of perfection online. Emphasize that authenticity involves accepting imperfections and that true connection is established through authenticity, not through an idealized image.

Reject unrealistic expectations online. Encourage people to resist the pressure to conform to unrealistic standards and to embrace their authentic identities and experiences.

Promote digital self-awareness. Encourage individuals to reflect on who they are online, what values they want to express, and how they can contribute to an authentic online environment.

Create an online culture that supports and respects authenticity. Establish community norms that promote mutual respect, understanding, and the appreciation of the diversity of experiences.

Promoting digital authenticity involves creating an online environment where people feel free to be themselves, share genuine experiences, and build meaningful connections. This culture of authenticity positively contributes to both individual and collective well-being in the digital world.

Ultimately, the relationship between technology and personal authenticity is dynamic and depends on how people choose to use the available technological tools. Awareness and balance are key to harnessing the benefits of technology while preserving and promoting authentic self-expression.

20.Technology and Longevity

The relationship between technology and longevity is a constantly evolving field of study that explores how digital tools can impact the quality of life and well-being in old age.

Technology facilitates social connection for older adults, allowing them to stay in touch with friends and family through video conferencing platforms, social networks, and online messaging. This can combat loneliness and improve the quality of life.

Social connection through technology is a fundamental aspect of improving the quality of life for older adults. Here are some additional points related to how technology facilitates social connection:

Technology enables older adults to feel included and actively participate in the digital society. They can join online groups, participate in thematic discussions, and contribute to virtual communities that share their interests.

Social media platforms and messaging apps allow older adults to reconnect with old friends and maintain relationships with geographically distant family members. This strengthens family and friendship bonds.

Technology enables participation in virtual events and activities. Older adults can attend online family gatherings, celebrate important events, and engage in classes, lectures, or recreational activities without leaving home.

Video conferences and virtual calls provide older adults with the opportunity to have face-to-face conversations with their loved ones, adding a visual and emotional dimension to communication, especially if physical distance is a challenge.

Technology allows for the creation and participation in online communities specifically for older adults. These digital spaces can be designed to address shared interests, concerns, and experiences, facilitating peer-to-peer connections.

Digital platforms provide access to shared photos, videos, and memories. This can be especially meaningful for older adults, giving them the chance to relive important moments and share stories with younger generations.

Technology enables the creation of online support groups where older adults can share their experiences, receive emotional support, and obtain useful information on specific topics of interest or concern.

Online games and recreational activities provide opportunities for fun and social interaction. Older adults can engage in virtual games, online challenges, and other recreational activities that foster connection and entertainment.

Technology can help reduce the feeling of isolation by offering a window to the outside world. Older adults can explore different cultures, news, and current events, contributing to a sense of connection with society at large.

Online emotional support platforms and chat services offer older adults a space to express their feelings, receive guidance, and connect with mental health professionals, which can be beneficial for their emotional well-being.

In summary, social connection through technology provides valuable opportunities for older adults to maintain meaningful relationships, participate in social activities, and overcome loneliness, thereby contributing to an improved quality of life.

Older individuals can use technology to easily access health information, online resources, and apps designed for health monitoring. This contributes to a proactive approach to personal care and the management of medical conditions.

Access to health information and resources through technology provides older individuals with valuable tools for personal care and effective management of medical conditions. Here are additional aspects related to this relationship:

Technology offers access to personalized health education resources. Older individuals can access specific information about their health conditions, treatments, and self-care practices, contributing to a more comprehensive understanding of their well-being.

Various mobile apps allow older adults to track their health. From monitoring blood pressure to controlling blood sugar, these apps provide simple tools for managing and recording relevant data for their well-being.

Apps and electronic reminders help older individuals manage their medication effectively. These reminders can be set to alert about medication intake, ensuring proper compliance.

Telemedicine allows older adults to attend virtual medical consultations, avoiding the need for physical travel. This is especially beneficial for those who may have difficulty accessing medical facilities regularly.

Physical activity tracking devices and apps give older individuals the ability to monitor their daily activity levels. This promotes a healthy lifestyle and can be particularly useful for those looking to stay active.

Online platforms facilitate communication with healthcare professionals. Through secure messages or virtual consultations, older adults can receive advice and answers to questions quickly and conveniently.

Apps and online resources provide information on nutrition and diet tailored to the needs of older individuals. This can assist in planning healthy meals and managing conditions related to nutrition.

Sleep monitoring devices and apps help assess sleep quality. For older adults, sleep is crucial for overall health, and having data on sleep patterns can be valuable for adjusting habits and improving the quality of rest.

Accessing information about wellness practices and disease prevention is essential. Technology allows older individuals to stay informed about healthy habits, recommended vaccinations, and disease prevention practices.

Technology facilitates communication between older adults, their caregivers, and family members. Sharing relevant health information, such as test results or changes in health status, can be faster and more efficient through digital platforms.

Access to health information and resources through technology significantly contributes to a proactive approach to personal care and effective management of medical conditions in older individuals, thus improving their quality of life.

Telemedicine allows older adults to receive remote medical care, especially beneficial for those with reduced mobility or living in remote areas. Virtual consultations can improve accessibility to healthcare.

Telemedicine and virtual consultations play a crucial role in enhancing accessibility and quality of healthcare for older adults. Here are additional aspects related to telemedicine:

Telemedicine enables older adults to access medical specialists without the need to travel long distances. This is especially beneficial for those living in areas where specialist availability may be limited.

Virtual consultations facilitate continuous monitoring of chronic conditions. Older adults can regularly communicate with their healthcare professionals to provide updates on their health progress and adjust the treatment plan as needed.

For those living in remote areas or with limited access to medical services, telemedicine reduces geographical barriers. This ensures that older adults receive quality medical care regardless of their location.

Virtual consultations offer greater convenience and safety, especially for those with reduced mobility. Older adults can receive medical care from the comfort of their homes, avoiding travels that may be challenging or risky.

After medical procedures or surgeries, telemedicine allows for postoperative follow-up through virtual consultations. This facilitates communication between the patient and healthcare professional to address questions, concerns, and assess recovery progress.

Older adults can have virtual consultations to discuss the results of medical tests. This enhances the patient's understanding of their health status and allows the healthcare professional to provide effective guidance and recommendations.

Telemedicine facilitates obtaining second medical opinions. Older adults can connect with experts from other locations for additional assessments of diagnoses and treatment plans.

Telemedicine is valuable for mental health consultations. Older adults can access mental health professionals without dealing with transportation barriers, especially important for addressing issues like depression and anxiety.

Through virtual consultations, healthcare professionals can provide education and advice on self-care practices, medications, and lifestyle changes, contributing to comprehensive health management for older adults.

Telemedicine promotes the continuity of care. Older adults can maintain an ongoing relationship with their healthcare professionals, essential for the long-term management of chronic conditions and the promotion of healthy aging.

In summary, telemedicine and virtual consultations are valuable tools that improve accessibility, convenience, and quality of healthcare for older adults, contributing to healthier and well-managed aging.

Applications and online games designed for cognitive stimulation can help maintain the mental health of older individuals. These programs offer mental exercises, puzzles, and activities designed to preserve and enhance brain function.

The use of applications and online games specifically designed for cognitive stimulation can be an effective strategy to maintain and improve the mental health of older individuals.

Applications and online games provide a wide variety of cognitive activities, ranging from puzzles and memory games to logical challenges. This allows older adults to choose activities that align with their preferences and specific areas of interest.

These programs are designed to address different cognitive skills, such as memory, attention, processing speed, and problem-solving. They provide comprehensive training to keep the brain active and agile.

Many applications and online games are adaptable to different skill levels. This ensures that older adults can start with manageable challenges and progress to more difficult levels as they improve their cognitive skills.

These programs often provide immediate feedback, allowing users to assess their performance and continue improving. Positive feedback reinforces the learning process and motivates older adults to engage continuously.

Games and activities designed for cognitive stimulation require focus and concentration. Regular practice can help strengthen older adults' ability to concentrate and pay attention, which is beneficial for overall mental health.

Many of these applications are accessible from mobile devices such as tablets and smartphones. This allows older adults to perform their cognitive exercises anytime and anywhere, making it easy to integrate into their daily routine.

Some programs allow collaboration between users and friendly competition. This fosters social interaction and connection among older individuals, adding a positive social element to cognitive stimulation.

Most of these applications are regularly updated, offering new challenges and activities. This prevents activities from becoming monotonous and gives older adults the opportunity to constantly explore new forms of cognitive stimulation.

Fun is an integral part of many of these applications. Incorporating playful and entertaining elements makes the process of cognitive stimulation more enjoyable, contributing to continuous engagement.

These applications can complement other cognitive activities, such as reading, writing, or participating in discussion groups. Together, these approaches can provide a comprehensive experience of cognitive stimulation.

Overall, the use of applications and online games designed for cognitive stimulation can be an effective and accessible tool to maintain mental health and promote cognitive well-being in older individuals.

Virtual assistants like Siri, Alexa, or Google Assistant can facilitate daily life by assisting with simple tasks, reminding about appointments, and providing useful information. Additionally, assistive technology, such as medical alert devices, can offer additional safety.

Virtual assistants and assistive technology play a crucial role in facilitating the daily lives of older individuals. Here are additional aspects related to these resources:

Virtual assistants like Siri, Alexa, or Google Assistant can automate everyday tasks, such as turning lights on and off, adjusting home temperature, playing music, or answering common questions. This simplifies daily life and improves accessibility.

These virtual assistants are efficient in reminding of medical appointments, important events, and medication schedules. Older adults can set personalized reminders, which is particularly beneficial for those with busy schedules or occasional forgetfulness.

The ability to ask questions and get quick answers through voice commands provides older individuals with instant access to useful information. This may include news, weather updates, cooking recipes, or personal interest data.

Virtual assistants allow older adults to communicate with family and friends through calls and voice messages without the need to use their hands. This is especially helpful for those who may have difficulties with more traditional devices.

Assistive technology, such as medical alert devices, can be essential in emergency situations. These devices can be linked to emergency response services or family members, ensuring prompt assistance in case of accidents or medical issues.

Assistive technology, particularly, is adapted to enhance accessibility for older individuals with disabilities. This may include devices responsive to voice commands, easy-to-use remote controls, or tactile technologies for those with visual impairments.

The ability to integrate with other smart devices in the home allows for centralized control. For example, virtual assistants can connect with smart thermostats, security cameras, and appliances, offering a safer and more comfortable environment.

Over time, virtual assistants can learn the personal preferences and routines of older adults. This enables greater customization in interactions, adapting to the specific needs of each individual.

Virtual assistants can provide support in daily activities such as creating shopping lists, searching for recipes, or planning recreational activities. This promotes independence and facilitates the completion of daily tasks.

Some virtual assistants are designed to provide cognitive stimulation. They can offer mental games, puzzles, or interactive stories, contributing to maintaining brain health.

In general, virtual assistants and assistive technology play a crucial role in improving the quality of life, safety, and independence of older individuals by facilitating various aspects of their daily lives.

There are applications designed to encourage physical activity and well-being in older adults. These applications can provide tailored exercise routines, track physical activity, and offer reminders to maintain a healthy lifestyle.

Exercise and wellness applications for older adults play a vital role in promoting physical activity and overall well-being. Here are some additional aspects related to these applications:

These applications offer exercise routines specifically designed for the needs and capabilities of older adults. They may include low-impact exercises, stretches, and activities aimed at improving strength, flexibility, and endurance.

Tracking features allow users to monitor their daily physical activity, including step count, distance traveled, and the amount of time spent on various activities. Visual feedback motivates maintaining an active lifestyle.

Some applications include customizable reminders to encourage regular physical activity. These reminders can be set to alert the need for active breaks or perform specific exercises throughout the day.

Integration with wearable devices enables real-time monitoring of physical activity. Wearables, such as fitness trackers, provide continuous data on steps taken, heart rate, and other relevant metrics.

Virtual classes and guided workouts can be part of these applications, offering a variety of exercises that can be done at home. This flexibility accommodates older adults who may prefer exercising in the comfort of their own space.

Nutrition tracking features may be included, allowing users to log their food intake and receive dietary recommendations. This contributes to a holistic approach to health and wellness.

Social features in some applications enable users to connect with friends or join virtual fitness communities. This adds a social component to physical activity, encouraging camaraderie and motivation.

Personalized recommendations based on user preferences and fitness levels can help tailor the exercise experience. This ensures that older adults can engage in activities that suit their abilities and preferences.

In summary, applications designed for exercise and well-being are valuable tools for promoting physical activity, maintaining a healthy lifestyle, and enhancing the overall well-being of older adults.

Applications often offer gradual training programs that allow older adults to start with manageable difficulty levels and gradually increase intensity as they improve their physical condition. To avoid monotony, these applications provide a variety of physical activities. They can include strength exercises, aerobic workouts, yoga, and tai chi, offering options to suit individual preferences and address different aspects of physical well-being. The applications are designed with user-friendly interfaces and provide clear instructions for exercises. This is especially beneficial to ensure that older adults perform movements safely and effectively.

Some applications can integrate with health tracking devices, such as smartwatches or activity bracelets, to provide more detailed and accurate monitoring of physical activity. These applications allow users to set personal goals in terms of physical activity and well-being. Tracking progress towards these goals can be a constant source of motivation.

To cater to different fitness levels, many applications offer short and personalized exercise sessions. This makes it easier to integrate physical activity into daily routines, even for those with busy schedules. In addition to exercise routines, some applications provide general wellness tips, such as guidelines for healthy eating, relaxation techniques, and strategies to improve sleep quality.

These exercise and wellness applications are valuable tools to help older adults maintain an active and healthy lifestyle, thus contributing to their physical and emotional well-being over time. Technology enables online learning, providing continuous educational opportunities for seniors. In addition, online communities and social networks can facilitate participation in cultural, educational, and recreational activities.

Online learning and community engagement through technology offer enriching opportunities for seniors. Here are some additional aspects related to these areas:

Online learning provides seniors with the opportunity to continue education and personal development. They can access courses on a wide range of topics, from art and history to science and technology, tailoring learning to their individual interests.

Online learning platforms offer flexibility in terms of schedules and location. Seniors can learn at their own pace and from the comfort of their homes, eliminating geographical and mobility barriers.

Online communities facilitate access to cultural resources, such as virtual museums, art exhibitions, and real-time cultural events. This allows seniors to explore and participate in cultural activities from anywhere.

Social media and online community platforms bring together people with similar interests. Seniors can join discussion groups, book clubs, or specific communities, making it easy to engage in activities they are passionate about.

Technology enables intergenerational mentoring and knowledge exchange. Seniors can share their experiences and knowledge while learning from younger generations in a collaborative digital environment.

Online platforms offer the opportunity to connect with experts and speakers from various fields. Seniors can participate in virtual conferences, webinars, and inspirational talks, expanding their horizons and perspectives.

Online communities allow active participation in discussions and community projects. This is especially valuable for those who want to contribute to society and connect with others with similar goals.

Online platforms offer virtual recreational activities, such as games, contests, and social events. These recreational elements encourage social interaction and enjoyment of recreational activities from the comfort of home.

Participating in online activities promotes the development of technological skills among seniors. This contributes to their digital autonomy and allows them to make the most of the opportunities offered by the digital environment.

Through online learning and community engagement, seniors can stimulate their minds, foster creativity, and stay mentally active, contributing to their overall well-being.

Together, online learning and community engagement through technology offer a comprehensive approach for seniors to continue learning, connect with others, and engage in activities that enrich their lives.

The integration of technology into smart homes, such as automation systems and assistive technology, can make the environment safer and more accessible for seniors, allowing them to live independently for longer.

The integration of technology into smart homes and the adoption of assistive technology offer significant benefits for seniors, enabling them to live more safely and independently. Here are some additional aspects related to this topic:

Technology in smart homes includes automation systems that allow seniors to control devices and systems easily. This may include lights, thermostats, door locks, and appliances, improving comfort and efficiency.

Assistive technology can incorporate sensors and devices to detect falls and emergency situations. These systems can send automatic alerts to family members or assistance services in case of a fall or medical event.

Home health monitoring devices allow seniors to track their vital signs and medical conditions from home. This may include blood pressure monitors, glucose meters, and other devices that facilitate health management.

Voice systems, such as Alexa or Google Assistant, allow seniors to control devices and access information through voice commands. This simplifies technological interactions and makes daily tasks more accessible.

Assistive technology integration may include automatic reminders for medications, medical appointments, and other daily tasks. This helps seniors stay organized and adhere to their routines effectively.

The installation of security cameras and surveillance systems in smart homes provides an additional layer of security. Seniors can monitor their home and receive alerts about unusual activities, enhancing peace of mind and safety.

Assistive technology focuses on universal accessibility, making homes more friendly for seniors and those with disabilities. This may include adjustments in switch height, easy-to-use faucets, and other elements that enhance mobility.

Technology in smart homes allows for personalized adaptations based on individual needs. For example, lighting can be automatically adjusted according to the preferences and needs of seniors, improving visual comfort.

Assistive technology can integrate with medical devices, allowing monitoring data to be effectively shared with healthcare professionals. This facilitates remote monitoring and management of medical conditions.

Overall, the integration of technology into smart homes and assistive technology provides strong support for seniors to live independently for longer, improving their quality of life and autonomy.

In summary, the implementation of technology in smart homes and assistive technology significantly contributes to the safety and independence of seniors, allowing them to enjoy a more accessible home environment tailored to their needs.

Security and surveillance systems based on technology can provide peace of mind to seniors and their families. These include security cameras, motion sensors, and automatic alerts in case of emergencies.

The implementation of technology-based security and surveillance systems offers peace of mind to seniors and their families by providing a safer environment. Here are some additional aspects related to security and surveillance for seniors:

Strategically installed security cameras in the home allow seniors to monitor specific areas and have a clear view of what is happening in real-time. This can be particularly useful for those living alone.

Motion sensors can detect unusual activities or movement patterns in the home. By integrating these sensors with security systems, automatic alerts can be sent in case of suspicious or unplanned movements.

Alarm systems can include door and window sensors, smoke detectors, and carbon monoxide detectors, among others. These systems alert residents and authorities in case of emergency situations, such as break-ins or fires.

Technology allows remote monitoring of the home through mobile devices. Seniors and their families can access cameras and receive security notifications, providing an additional layer of control and peace of mind.

Setting up automatic alerts allows receiving real-time notifications about security events. This may include alerts for motion, open doors, or changes in normal activity in the home.

Integrating security systems with assistive devices, such as emergency buttons or medical response systems, ensures a quick response in critical situations, such as falls or medical emergencies.

Security technology allows family members and caregivers to access surveillance information. This is especially valuable for those who want to monitor the well-being of seniors from a distance.

Security systems may include access control options, such as smart locks. This allows seniors to manage who can access their home, improving security and privacy.

Automated security lighting can be linked to motion detection systems. This ensures that specific areas are automatically illuminated when activity is detected, deterring potential intruders and enhancing security.

It is essential to ensure privacy and data protection when implementing security systems. Seniors and their families should be aware of privacy settings and have control over who accesses the collected information.

The combination of security cameras, sensors, and technology-based alarm systems provides an additional layer of security for seniors, allowing them to live more independently and providing peace of mind to their loved ones.

Technology that focuses on accessibility and inclusive design ensures that digital solutions are usable by all people, regardless of their age or abilities. This contributes to a more equitable technological experience.

Accessibility and inclusive design in technology are crucial to ensuring that digital solutions are usable by all people, regardless of their age or abilities. Here are some additional aspects related to accessibility and inclusive design:

Accessible technology is characterized by intuitive interfaces that allow easy navigation. This benefits older adults by making applications and devices easier to understand and use.

Options to adjust font size and contrast are essential to ensure that information on digital screens is readable for all people, including those with visual difficulties or presbyopia.

Inclusive technological solutions are compatible with assistive technologies, such as screen readers, special keyboards, and alternative input devices. This allows people with disabilities to fully interact with technology.

Inclusive design involves testing with diverse users, including older adults, to identify potential barriers and make adjustments that improve accessibility. This ensures that technology is suitable for a wide range of users.

Audio description and subtitle features are essential to make multimedia content accessible. This benefits people with visual or hearing impairments, as well as those who prefer consuming content silently or with additional descriptions.

Ergonomics and adapted design are key components of inclusive design. This involves considering the ease of use of devices and applications, taking into account the physical and cognitive abilities of users.

Providing feedback and contextual help enhances the user experience for older adults. Clear instructions and assistance at the right time contribute to a smoother interaction with technology.

Inclusive technology often includes voice navigation capabilities and voice commands. This benefits those who may struggle with tactile interaction or prefer using voice as their primary means of communication.

Inclusive design considers cognitive limitations and provides simplified interfaces. This helps older adults who may experience cognitive difficulties when using technology.

Commitment to continuous updates and improvements reflects an inclusive approach. This allows addressing user feedback and adapting to changing needs and emerging technologies.

Inclusive design in technology is essential to promote equity in access and use of digital devices and applications. It ensures that solutions are accessible to a wide range of users, including older adults, enhancing their technological experience.

Although technology offers many positive opportunities, it is important to address challenges such as the digital divide and ensure that solutions are intuitive and accessible for all ages. Overall, technology can play a crucial role in improving the quality of life and well-being of older individuals.

www.ingramcontent.com/pod-product-compliance
Lightning Source LLC
Chambersburg PA
CBHW072358290526
45794CB00001B/104